Michel the Giant

Tété-Michel Kpomassie was born in 1941 in Togo. At the age of sixteen he ran away from his traditional village and embarked on a twelve-year journey through West Africa and Europe to reach Greenland. Once there, he immersed himself in the life and customs of the local people and the landscapes that had captured his imagination from afar. On his return, he wrote about his travels under the working title of *Mikilissuaq* (*Michel the Giant*), the nickname he had been given by the children he met in Greenland. The resulting travelogue, published in French as *L'Africain du Groenland* in 1977, was awarded the Prix Littéraire Francophone International and shortlisted for the Thomas Cook Travel Book Award.

TÉTÉ-MICHEL KPOMASSIE

Michel the Giant

An African in Greenland

Translated by James Kirkup
Afterword translated by Ros Schwartz

PENGUIN BOOKS

PENGUIN CLASSICS

UK | USA | Canada | Ireland | Australia
India | New Zealand | South Africa

Penguin Books is part of the Penguin Random House group of companies
whose addresses can be found at global.penguinrandomhouse.com

Penguin
Random House
UK

First published in France under the title *L'Africain du Groenland* by Flammarion 1981
This translation first published under the title *An African in Greenland* in the USA
by Harcourt Brace Jovanovich 1983
First published in Great Britain by Martin, Secker & Warburg 1983
Published under the present title and with minor revisions and a new
Afterword in Penguin Classics 2022
005

Copyright © Flammarion, 1981
Translation copyright © Harcourt, Inc., and Martin Secker & Warburg, Ltd, 1983
Copyright © Tété-Michel Kpomassie, 2022
Afterword translation copyright © Ros Schwartz, 2022

Set in 11.25/14pt Dante MT Std
Typeset by Jouve (UK), Milton Keynes
Printed and bound in Great Britain by Clays Ltd, Elcograf S.p.A.

The authorized representative in the EEA is Penguin Random House Ireland,
Morrison Chambers, 32 Nassau Street, Dublin D02 YH68

A CIP catalogue record for this book is available from the British Library

ISBN: 978-0-241-55453-1

www.greenpenguin.co.uk

For Jean Callault

Contents

Contents

PART I

The Python God

The Snake in the Coconut Tree

'Not awake yet, is he?' Uncle asked contemptuously.

He spoke softly, making visible effort not to raise his voice either to control his anger or so as not to disturb those who were still sleeping in the neighbouring huts.

'Not yet,' my brother Tété replied. 'What a job it is to wake him! May the gods forgive me for what I'm about to say, but I think if a thunderbolt struck the roof, he'd still go on sleeping! I shook him and smacked his bottom, but he never even stirred!'

'Is he breathing?' Uncle asked.

'Breathing? Like a blacksmith's bellows!'

'So he's only pretending to sleep. He just wants to spend the day alone at home doing nothing. Like a big fat lizard, the lazybones!'

There was a moment's silence, then Uncle said:

'Splash cold water over him.'

A shiver ran through my body when Uncle uttered those words, but still I made no movement. I was rolled up in my loin-cloth, arms folded under my head. I squinted at my brother Tété. By the dim glow of the palm-oil lamp I could see him moving towards a corner of the room, behind the door, where we put drinking water to cool in an earthenware jar called a *canari*. He picked up the gourd which lay on the *canari's* wooden lid, filled it brimful of water, took it in both hands, and crept towards me. I leapt to my feet, hugging the loincloth to my body, and implored my brother not to throw the water over me. Reluctantly, he put the gourd on the ground, though not before dipping his fingers in it and flicking a few drops of water in my face.

Under the stern eye of my uncle, still standing in the doorway, I quickly rolled up my sleeping mat and leaned it against the wooden wall at the far end of the room. Then I started looking for my khaki shorts. They were on the beaten earth floor of the hut, in the corner where I had thrown them before I went to sleep: we were forbidden to sleep in anything except a loincloth. After putting on my shorts, I picked up the loincloth I had used as a cover during the night. It was an indigo loincloth, a colour I loved. The material, a beautiful cotton twill, had been dyed by one of my *navi*, the wives whom my father had married after my mother. Now I had to tie the cloth around my waist, or else wrap my whole body in it against the morning chill. To do this, we stretch the loincloth lengthwise behind our backs, holding it by the two upper corners, which we then cross over the chest and tie around our necks. It took me some time, because I was still only half awake. I kept yawning and scratching myself.

'Get a move on!' yelled my uncle. 'Everyone's ready but you. It'll soon be daylight!'

Uncle was exaggerating. The cock had just crowed for the second time.

We used to set off like this, very early in the morning, to gather coconut tree fibres and branches. We used the fibres to make straw mats for sale. That morning we had got up earlier than usual. In fact, the night before, we had decided to leave before dawn, hoping to return at noon and spend the rest of the day weaving our mats. It was the school holidays. We were forbidden to climb the coconut trees near the houses and inside the courtyards, to avoid accidents if people (especially children) were to pass under the trees just as we were cutting down a bunch of coconuts. Despite this ban, we had already collected all the good nuts from the trees in our neighbourhood. Now we had to go farther afield to find new ones.

That morning, I didn't feel like going to the coconut plantation; I had a foreboding. Uncle wasn't altogether wrong to call me a lazybones, but a few hours later he would regret having forced me to go with them.

I tied my loincloth around my neck and we were finally on our way. The courtyard was in darkness. On our left, a large building roofed with corrugated iron occupied the centre, facing the entrance; my father lived there, and he was still asleep. In front of us, along the wall, was a row of huts belonging to the five wives he had at that time. These huts were made of unbaked bricks and roofed with straw. They formed a linked series of individual single rooms without communicating doors but very spacious, for each of our mothers slept there with her younger children of both sexes: on one side, the smaller boys, who were not considered real men and so not entitled to separate huts in the big courtyard, and on the other side the girls of all ages. The girls would get married some day and leave the family, so they lived there only temporarily. Father seldom entered these huts, and we had never known him to sleep in any of them. Instead, for one month each wife would leave her hut in the evening and live in intimacy with our father in the big central building. The wife who shared his bed was the one who made his meals for that month. As a rule, she cooked two separate meals: one, dainty and delicious, for Father, which required endless time spent preparing various little appetizing dishes; the other, less elaborate, for her children, whose food she nearly always shared, while Father took his meals alone.

This monthly selection of a wife was not done by lot or by my father's choice, but was in order of seniority, from the first to the last wife in fixed rotation, so as to avoid threats of divorce.

Like it or not, each of the wives was periodically omitted and had to absent herself from felicity because of our father's rigorous observance of certain traditional prohibitions. The wife whose turn it was must not be in the process of being 'visited by the moon', which causes menstruation, and must not be nursing a child. Any wife in either of these states was temporarily impure in our father's eyes, and he would not touch food prepared by her or share his roof with her. (The children, both boys and girls, were not affected by this ban.)

When we got up, none of the wives was yet out of bed; no sound came from that side of the courtyard, though behind us, through the closed doors, we could hear our elder brothers snoring. We stepped over the sheep and crossed the vast courtyard without a sound, the soft sand muffling our footsteps.

When we reached the well, whose cemented rim formed a darker mass in the blackness between our mothers' huts and our father's house, I drew two buckets of water which I poured into a larger bucket, where we dipped our hands to wash our faces and rinse our mouths. Just then, Kunugnan (Innocent), our household dog, left the threshold of my father's hut, approached us and, wagging his tail, tried to lick our wet faces. After making sure to empty our washing water on the sandy ground, we left the compound. Kunugnan followed us.

Soon we had passed the last houses in Kpéhénou, one of the outlying suburbs of Lomé, and set out along a narrow footpath through the bush. The great feather dusters of the coconut trees, now more numerous, were swaying above our heads like gigantic parasols whose outlines stood out in infinite recession against a pale sky that was slowly growing brighter. Not far away, a dark blue sweep of sea was sparkling in the first rays of the rising sun. A warm breeze played around us, slowly swaying the ribbed branches of the coconut palms and dislodging big drops of dew that spattered our bare shoulders, making us shiver.

I was the youngest of the three, so according to custom I was walking in front. Behind me came my 'revered big brother' Tété, five months older than me. He was the first son of my father's second wife. Our young uncle, Ahouanssou, brought up the rear. He was almost as old as our eldest brother, but none of us, not even my big brother, was allowed to address him by name. We simply called him *Atavi* (paternal uncle); he was one of our father's cousins. Nor were we allowed to address our older brothers – even if they were only one week older than us – without using the title *Fofo* before their names. This means 'revered big brother'. Even twins are not exempt from this rule in our traditionalist families, though for them the order is reversed: the elder of the twins is the one who is born last. We explain this by saying: 'It is the last-born twin who ordered his little brother to go out first and find out if the world was worth looking at. And depending on the newborn child's first cries and what he says, the big brother comes out either alive or stillborn.'

Togolese fathers cannot conceive of any community, however small, without authority, without one or several heads who issue the orders; the absence of hierarchy is unthinkable, even among brothers! As the sixth of my father's many children, I had to obey without a murmur my four elder brothers and one elder sister, not to mention this very distant uncle. Of course the rest of my brothers and sisters had to show me the same respect and obedience. So if an order was given for some tiresome job, the elder brothers could get out of it by foisting it on the younger ones. The youngest son of the family, crushed flat at the base of a pyramid whose summit was our father, was the one to be pitied most.

*

We walked on without a word, chewing our toothpicks.

Custom did not lay down the position Kunugnan should occupy on that narrow path, so he often amused himself by scampering off. He would gambol through the thickets, his body drenched with heavy drops of dew, sniffing out dead lizards and pausing outside rat holes which had been taken over by snakes.

The sight of a dead lizard suddenly recalled a fading memory. I must have been about twelve or thirteen, for I had long been circumcised. We were hunting a kind of lizard with a pink, iridescent skin. It is hard to kill them on the ground because of their astonishing speed, but sometimes, unfortunately for them, they try to escape by climbing a coconut palm. We call them, in the Mina language, *adimbolo* or else *adambolo*. To hunt them we would use a long whip of fine, plaited wires. When we spied a lizard clinging to the trunk of a coconut palm, we crept around to the side opposite the animal, then cautiously approached the tree, the whip concealed behind our backs, and placed the index finger of the other hand against the tree trunk. We believed these creatures were deaf, but all the same we never spoke or made any noise while approaching for the kill. One of our brothers would stand some distance away, making signals with his head, not with his hands, to let us know if the position of that finger was exactly opposite the centre of the body of the wary reptile crouched stock-still on the other side of the trunk. 'No, not there . . . a bit higher . . .' the brother would signal, slightly lifting his chin. 'A bit lower . . .' he would continue, with a slight downward movement. 'Yes, that's it,' he would finally grin. Then the hunter would silently take a step or two back, gather all his strength, and *whack*! – give the coconut tree a vigorous lash while at the same time pulling on the whip, which zipped around the trunk. The lizard was nearly always killed on the spot.

When we had killed a dozen or so in this way, we would hide among the bushes, so as not to be interrupted by the adults. There we gutted the lizards with sharp stones or with shards of broken bottles. Then came an operation of great importance in our adolescent lives. We arranged the corpses on a piece of broken earthenware or bit of iron sheeting that we placed – unknown to our parents, and especially to our mothers and sisters – on the roof of a shed. We kept a careful watch on these lizards. When the weather was rainy, we smuggled them into our rooms, not minding having to live with the smell all night. Finally, after several days on the roof under the burning sun, the lizards would sweat a tiny quantity of melted grease that we carefully collected. After this, the carcasses were thrown away or sometimes even buried, so as not to arouse suspicion.

The ceremony which then took place was conducted by our *Fofogan,* the first son of our father's first wife, therefore the eldest son in the entire family. We held it in the middle of the night when everybody, including our father, was in bed. Then, in one of our rooms we would form a semicircle, standing on the beaten earth floor or squatting on a mat. We would remove our loincloths. *Fofogan,* also naked, would dip his fingers in the lizard grease and spread it on his penis in layer upon thin layer. Then it was our turn, according to age. The scene took place by the flickering light of a little palm-oil lamp. The aim of this operation was to make the penis longer and thicker, and to prolong erection, and those, we believed, were only a few of the grease's many virtues. We didn't wash ourselves for three days after applying this grease, so that, we claimed, it could work more effectively by getting right under the skin. For us, this avoidance of washing was the hardest part of the operation, for custom required us to wash three times a day, before each meal. So when we carried fresh water into the wash place, we would splash and trickle it all around, in a pretence of washing. We repeated the lizard grease ceremony time after time: some-times, the lizards yielded so little grease that we were forced to use the carcasses. However, three days after each ceremony, we went back to regular washing in the traditional way. Then we started all over again. Finally, in order to test the efficacy of the grease, we would try our luck with the neighbourhood girls.

Then our *Fofogan* would gather us together once more in great secrecy, and ask each one of us who had tried the ultimate experience:

'How many times?'

'Three times,' some would say.

'Five!' boasted others.

'But it should have been *seven* times!' he would cry in con-sternation, and the ones who scored too low had to go through the whole procedure again.

Once more, the lizard war began.

Kunugnan came running out of a thicket towards us with a liz-ard in his jaws: he laid it at our feet, but it was not the right kind. We continued along the path, each sunk in his own thoughts. The morning sun was rising; with each moment the sand was growing warm, and we knew that towards eleven o'clock, on our return, the path would be burning hot under our bare feet. So we hurried. Our machetes, whose notched blades were thrust through our belts, slapped against our thighs. Soon we had reached our destination.

Whenever we started to cut branches in the coconut grove without first going to say good morning to the watchman, he would start giving us trouble as soon as he spotted us in the trees: threatening us with his machete, he would order us to come straight down, then chase us away and confiscate the branches we had cut. But if we went to pay him our respects as soon as we arrived, he would leave us in peace.

Before making our way to the watchman's house, I handed three five-cent coins to my uncle, as did my brother, to which my uncle added his own share. Then we took a shortcut, walking on the grass. Uncle was in front, followed by my brother, while I walked behind them now, for it was up to the eldest to approach the watchman and do the talking.

This watchman lived alone among the coconut trees, in a straw hut surrounded by a fence; and there were no other houses for several hundred metres around. The tight-woven branches of the fence were firmly tied to thick stalks stuck into the ground. These stalks would begin to grow again, forming a dense hedge above the fence that half-concealed the conical roof of the hut in the middle. Mountains of copra for the manufacture of coconut oil were piled up in front of the house, near the water trough, and some of them had begun to sprout, too. Hens scratched about in the courtyard.

Uncle clapped his hands to announce our approach. A moment later, the watchman appeared on the threshold of his hut and motioned to us to wait. He crossed the yard, making his fine white slippers slap against his bare heels as he walked. He wore a long blue tunic, and his shaven head was covered by a conical cap pulled down to just above his jet-black, slanting eyes. His hands were wrinkled, and the veins stood out like gnarled, spreading roots. His weasel face, ending in a little pointed beard, seemed very funny to us. He was a Peuhl herdsman – a different tribe from ours – and didn't speak our language well. Although he might have been about our father's age, we didn't call him by the respectable title of *Ata*, or papa, as we did other adults, for he was a 'foreigner'. Instead, we called him *Yessuvi* (little Jesus) because of his goatee beard and the solitary life he led. But as he was a Muslim, this nickname infuriated him. He had no wife, and on this score we badgered him with all kinds of rude and even cruel jokes. As he kept cattle, we would ask: 'Hey, *Yessuvi*, is it because you prefer doing it to your cows?' Then the culprits would take to their heels. Fortunately for us, he did not always recognize his persecutors when we returned to the coconut plantation. So he greeted us with a broad smile, revealing two rows of disgusting teeth yellowed by cola-chewing.

We didn't have time to tease poor old *Yessuvi* or to play any tricks on him, so after the usual greetings *Atavi* handed him the forty-five cents. 'That's very kind of you,' said the watchman, closing his fingers on the coins. 'You're good boys, you are. Go and pick the nuts, but don't waste the new ones.'

Then he turned round and left us.

That immense coconut plantation, called the 'Pa' or 'Papa de Suza' plantation after a Togolese VIP, covers all the southern region, from the Ghanaian frontier to the borders of Dahomey (Benin) – a distance of fifty kilometres. The trees stand one behind the other in perfectly straight lines, which shows that they did not spring up naturally. Often I would count twelve paces from one tree to the next and see only a tiny crack of blue sky between the spreading branches overhead. Sometimes you would come across a tree still solid enough at its base, but dead at the top – it had been struck by lightning. The weird impression made by these leafless, blasted trees was similar to what one might feel on seeing lepers, who have lost their fingers, in the middle of a crowd of healthy people. Nobody touched these blasted trees; they were left to rot and crumble on their own. After a few years, when nothing remained but roots and the stump of a trunk, the watchman dared to approach the spot and replace them with saplings taken from the piles of copra and cultivated in a kind of nursery next to his house.

I've no idea why, after leaving *Yessuvi* that morning, we headed westward into the plantation, when we usually went north. We moved away from the sea, which a narrow strip of beach separated from the first row of coconut palms, whose slender trunks were green from the spray.

The track we were following grew narrower and narrower. Tall bushes came right up to my shoulders. Some leaves, when they touched our bodies, gave us a violent itch and made us scratch furiously. Birds with brilliant plumage flew off at our approach. The undergrowth to our left and right was a mass of all sorts of shrubs. There were some *djémakpan* (salt-leaf plants), and an abundance of the curious fern that in the Mina tongue we call *miongui-miongui* or, in other regions, *mianta-mianta*, both expressions having the same meaning which, freely translated, might be rendered as 'modesty', because this fern reacts quickly to the slightest contact. The great gales of the bush, as they batter the fern from all angles, have no effect upon it, but it closes up its leaves as soon as you touch it with your fingers. The leaves on the branches which have felt the contact react immediately by standing up symmetrically, two by two, flattening themselves against one another. They cling timidly together along the branch for about five minutes before beginning to open slowly again, thinking the intruder far away. As soon as you touch them again, the same thing happens.

The track finally led us to a wide open space where the grass had been cleared, and Uncle decided that we would make this our meeting place because now we were to separate, each going his own way to look for branches that he would bring back to this spot in small bundles of a dozen or so. This precaution allowed us to keep in contact with one another on that immense plantation. If one of us took too long to report back, we would go and see if anything had happened to him.

Our loincloths usually got in the way when we were shinning up the trees, so we preferred to leave them behind at our meeting place, taking care to roll them up neatly and place a small bunch of grass or a handful of sand on top of them. This way anyone passing by in our absence would realize that the loincloths had not just been discarded by their owners. Like any other Mina, I would never take any object marked like this because the tuft of grass and the handful of sand bring bad luck to anyone who steals the object hidden underneath. We are brought up in the belief that anyone appropriating an article covered with such signs risks the vengeance of Hêviesso, the lightning god, or of Sakpatê, the earth goddess – represented here by the tuft of grass or the handful of sand – whose punishment comes in the form of smallpox (unfortunately, she forgets that it's contagious).

My brother was the first to leave. A few seconds later I caught sight of him in the distance. He was halfway up a coconut palm. From where I was standing he looked like a giant ant, and he was climbing steadily. Soon he had reached the top and merged into the great fan of leaves. *Atavi* had left, too, but the high grass round the clearing kept me from seeing which way he had gone.

Left alone, I in turn moved towards a tall, slender coconut tree about twenty metres high and only a few paces away from where we had deposited our loincloths. It was enveloped in creepers. A shadowy fear took hold of me as I looked up at its great tufted crest. Though the tree was big and solid at the base, near the crown it was incredibly thin, swaying precariously in the wind. But it was loaded with ripe coconuts, so I quickly shrugged off the apprehension inspired by the creepers and the fragility of the upper trunk, till I had only one thing in mind – to send all those coconuts crashing to the ground! I even felt glad that neither my uncle nor my brother had been the first to notice this special sort of greasy pole with prizes at the top.

My bare foot was already set on the broad swelling formed at the base by exposed roots. My arms embraced the trunk in a powerful squeeze, I spread my legs out, flexed my knees to raise my feet and clamp the tree between my thighs, then relaxed the grip of my arms and heaved myself farther up by shoving with my feet. In this way I used my feet and arms in turn to hoist myself rhythmically up the tree. Clambering without a rope up coconut trees, which are nearly always smooth at the base but rough from the middle up, was an almost daily exercise for us boys, and I had no trouble doing it, though it often left me with a few grazes on my chest. But those very scratches made us proud! Often the scar tissue would open again as we were climbing.

I had reached the thinnest part of the trunk, just under the first dry, reddish branches hanging around my head, and I stretched out my legs horizontally, then crossed them to squeeze the trunk firmly between my thighs and gain a solid grip, enabling me to free my arms. Taking hold of a dry branch, I gave it a tug: it came away easily and I was showered with dusty debris. This often contains small creatures like centipedes (whose bite is painful, sometimes dangerous) and scorpions. It was to prevent these nasty little creatures from slipping into our pants that before climbing we would tie round our waists a belt or even a piece of rope that also served to hold our machetes. I shut my eyes so as not to be blinded by dust, and quickly ran my hands all over my body to dislodge those little pests, if there were any. An instant later, I heard the branch land on the ground.

After pulling off two or three more branches, all dry as tinder, and cutting off some yellow ones, I found myself high up in the green branches of the palm. These are tougher than the others and cling strongly to the trunk, from which they still draw sap. They spring vigorously skyward before bending halfway down to form that familiar graceful curve. Each one forms a large cavity at its base, where it joins the trunk, and these cavities collect rainwater. However, as the branches of a coconut palm grow one above the other, the highest shelter the ones below, which get less rain. Some species of birds make their nests there.

I grabbed the green branches one after another, shifted my feet on the stem, and hoisted myself to the very top of the tree, where I settled fairly comfortably, shivering in the wind. All the bunches of coconuts were now clustered underneath my feet, so that by giving two or three thumps with my heel, I could easily shake a whole bunch loose. But first I wanted to reward myself for the effort I had made in getting up there, so I sliced the top off a big, tender nut and drank all its milk, burping beatifically at intervals, then got rid of the empty shell by dropping it to the ground. It was just at that moment that, glancing sideways, I saw right next to me the gleaming neck of a snake that was furiously swaying its scaly head while its long, thin, forked tongue kept flickering nervously in and out. The rest of its body was coiled in the cavity at the base of a big branch, and its thick loops lay across a prodigious mass of eggs: some of these had already hatched, and the reptile came no closer, probably for fear of dislodging the baby snakes that twined around its body.

The moment I set eyes on that horrible creature and her young, I was overwhelmed by terror. I couldn't kill it with my machete, because we had often been warned not to cut a snake in two with any kind of blade, as we might be struck by the severed head jerking in atrocious convulsions and its jaws might fasten on some part of our body. The only way to kill it was with a stick, and I didn't have one. In any case, the paralysing fright that took hold of me made the machete slip from my hands.

From that great height, I didn't dare drop to the ground. Taking a good grip on two solid branches, in seconds I was clutching the trunk again, meaning to slide down to the ground. But the snake was quicker than me; she shook the baby snakes off her sinuous body and started unwinding down the trunk towards my forehead. I was in such a panic that I lost all sense of danger and was instinctively driven to use my own bare hand as a weapon. The reptile, which may only have been trying to see me off, swarmed down after me in my rapid descent. As the snake slid towards me like a long ripple of water, I could see that terrible white throat raised slightly away from the rugged trunk of the coconut palm, while the rest of its long body hugged the tree. I don't know exactly when the edge of my right hand struck that slack body, but the blow shook it loose. Wriggling, it slid over my hair, then down my spine, spun in the air like a great lasso and thumped down on the sand below. How lucky I was to have climbed up without loincloth or shirt, or the reptile might have lodged there. In a flash, I remembered the premonition I had had at dawn, when I hadn't wanted to get up, and my apprehension before scaling the tree.

I felt relieved to be rid of the snake, but a few moments of violent aftershock made me tremble all over. Then I pulled myself together and went on down as quickly as I could, though fearful of not having the strength to reach the ground. Lost in a sort of daze, I happened to glance downward and was stunned by an unexpected sight: the snake, having apparently wasted no time lying on the sand, had started to slide back up the tree towards her young. I couldn't believe my eyes. How was I to get down? Not for anything in the world would I face another meeting. I had no desire to see it sliding over my body again or fastening its fangs in one of my bare feet. So halfway down the coconut tree, I sprang into space. A drop of about ten metres.

I hit the sand with an impact that shook me to the bone. There was a stabbing pain, a terrible crash, a sort of lightning flash, then total blackness. I made a superhuman effort to drag myself along for a couple of metres or so, digging my elbows in the sand, wriggling, straining, trying to get up, but in vain. I blacked out.

My brother and my uncle told me, much later, that I was *kou pégni* (half dead). Alerted by my shouts when I was struggling with the snake, they had caught a distant sight of it slithering over my head. They had come running up with sticks, but too late.

2

The Sacred Forest

I don't know how long I remained unconscious. Hours, perhaps days later, I came to. Sweat was pouring down my face. Painfully I turned my head and gazed about me. I was lying on a mat spread over a smooth, cemented surface which was nothing like the uneven beaten earth of our rooms. I let my eyes wander over the walls and the objects in the room, and realized that I knew where I was – in my father's living room. I recognized the voice of a neighbour, a friend of my father's, saying: 'It was lucky he passed out right away, or else the fear and commotion would have carried the venom straight to his heart.' Besides my father and his friend, there were five or six people bending over me; I couldn't recognize all the faces. I closed my eyes again and once more lost consciousness, before I could answer a single question. It was not until much later that I discovered what had happened when they brought me home.

They had inspected my hair carefully to try to find the marks of the snake's fangs, for my uncle and brother had said I was bound to have been bitten on the head, but they could find nothing. Then they had turned me this way and that, but among the bruises that covered my body like a rotten mango, they had not been able to discover the fatal wounds they were looking for, those two cruel little punctures that a snake usually leaves on its victim's body. Nevertheless, my father had sucked the blood out of all suspicious marks. To do so, he filled his mouth with *sodabi*, distilled palm wine. Perhaps it was the burning sensation of this alcohol on my wounds that brought me around from my second fainting spell. When at last I was able to answer questions, my father asked me again and again:

'Were you bitten?'

Painfully I shook my head to say: 'No.'

'What?' he cried. 'The snake didn't bite you? Are you quite sure? Tell me quickly!'

'No. It slid over my head, but it didn't hurt me.'

At these words, there was a great stir all around me. My father was alarmed. I soon learned why when I felt a bitter taste in my mouth. While I was unconscious, he had forced me to swallow eight pills made of roots and toad venom. It was an antidote, and the eight pills amounted to a very strong dose, but as I hadn't been bitten after all, my father realized that his homemade antidote, far from bringing relief, was slowly poisoning me. That was why his face showed such anxiety. He had me drink plenty of water, then stuck two fingers down my throat and made me vomit copiously.

I had broken nothing, but after two days I still could not get up. I lay there in the living room, where they had covered me with my father's ceremonial loincloth, woven and decorated with bands of colours all broken up like the design of a parquet floor. It was a loincloth of the kind known as *kenté*. Some thinner loincloths were spread on the mat, and the wives had rolled up others and put them under my head as a pillow, taking care to leave out any that were plain white, a symbol of death, for the dead are buried in white cotton loincloths.

On the very first evening after my accident, fever had set in. I was delirious, and my nightmares swarmed with snakes. Everybody was worried, yet they didn't take me to the hospital, either because it didn't occur to them or because tradition dictated otherwise. My father treated me with two or three different herbal potions, which I drank in small doses, and twice a day – in the morning at about eleven and in the evening before sunset – they took me out to bathe me. Rather than take me to the nearby yard, where the boys and girls who still lacked hair in their armpits and at the groin could bathe together, they walked me slowly to the wash place at the other end of the patio, because I was already adolescent.

On the evening of the second day, my condition worsened. At the peak of the fever my delirium increased, with prolonged intervals of lethargy.

At nightfall my father's first wife – whom I addressed according to tradition by the respectful title of *Nagan*, because she had been married before my own mother – entered the living room to bring me food. She had brought some *akassa*, a porridge with a maize flour base. *Nagan*, bending over me, was struck by the fixity of my gaze. My eyes were dim and I was staring into space, with no feeling in my limbs. My breathing was slow. *Nagan* put down the calabash of *akassa*, and waved her hand in front of my face: my eyes didn't follow it. Then she spoke my name: I didn't answer. She thought she was watching my last moments! That's how we generally await death, not clinging to life with frightful moans, but calmly and with resignation, much as we wait for a train or a 'bush taxi' in our country, indifferent to the time of departure or when we'll reach our destination.

Nagan called out in alarm to my father, who was sorting plants in the yard, in the big barn that we used as our kitchen.

'*Fofo*, come quickly!'

Father came running, a loincloth knotted around his hips. When he saw the state I was in, he sent two of my brothers to look for more plants, especially one root whose name I forget; then he sent two other brothers to fetch a white chicken. It appears that, while in that state of prostration, I had a brief moment of hallucination when I spoke a phrase that sick people close to death often pronounce in our land. Apparently I told them that I was standing beside a great river; drawn to the opposite bank by the beauty of its landscape, I was calling out for a boat to take me across. That signified that I wanted to pass from this life into the next, and my father said this would inevitably have happened if, during my delirium, I had taken my place in the dugout canoe I was asking for in my dream. As soon as I spoke these words, father dismissed his weeping wives and shut the door of the room. Only *Nagan* remained. My brothers hadn't been able to find any completely white chicken in the yard: the one they brought had a few black feathers at the tips of its wings.

'That's not what I need,' he told them. 'Run and get a completely white one from the neighbours!'

The brothers who had gone off into the bush returned half an hour later bringing plants they had collected, roots and all. Father cut off one of these roots, wiped it on his loincloth, then snapped it apart and dripped its milky sap into my nostrils. I felt a prickling sensation in my brain, and in a few moments sat up and sneezed violently. Soon afterwards my breathing went back to normal. Then they brought the white pullet.

Father sat me up and passed the chicken fourteen times around my head, seven times one way, seven times the other. Then he passed it as many times over my body, from head to foot and foot to head, letting its outspread wings brush my body. He was holding the chicken by its feet, head dangling. Then he cut its throat and poured all the hot blood over me. He skinned the bird without pulling out a single feather and placed it on its back, the stomach open and pulsing, in a big earthenware dish covered with a white cloth. Next, he rubbed my body all over with cowrie shells,* which were cold to the touch, then

---

* Shells used as money in Black Africa.

with shelled cola nuts, and placed everything in the dish beside the chicken. That night one of my brothers went outside to deposit this dish and its contents at a crossroads. No one must recover that beautiful dish or its contents: it was a *vossa*, a sacrificial offering. The chicken had exchanged its brief life for mine to let me live a few more moments, the time my father needed to finish the lengthy preparation of the plants he was sorting in the yard. Soon afterwards the fever dropped, and I fell into a deep sleep.

When I wakened during the night, I heard the sound of arguing on the veranda. *Nagan* was saying:

'You shouldn't leave the child like that, trying to cure him yourself, even if you are a good healer!'

It was the first time I had heard her talking to my father in that tone of voice.

'I'm just a poor woman,' she went on firmly, 'but I'll tell you what I think. Before stuffing a sick boy with roots and herbs, first find out the cause of his sickness. Do we even know why the snake attacked him up in the tree? Who knows whether he hasn't perhaps offended the ancestral spirits or done some wrong to snakes. Terrible accidents never happen by chance – there's always some hidden cause.'

There was a silence. I could well imagine my father looking askance at his wife, surprised by the assurance in her voice. His condescending answer astounded me:

'So what do you suggest? Send him to the hospital, where the doctors know nothing of our customs? Do they even know how we prepare our sick to face death when they're dying?'

'No, *Fofo*, I'm not saying we should take him to the hospital, but to Bè, in the sacred forest, to be cured by the followers of the snake cult. Only their *bokonon* can get to the root of the accident, discover its meaning, and perform the right sacrifices and the proper cure. You can't do one without the other!'

The *bokonon* are priests who are in touch with divinities. Father, who was himself a *bokonon*, possessed a vast knowledge of plants, their virtues, and the illnesses they can cure: he knew how to blend and administer them. He worshipped and consulted a certain number of divinities. But, not being initiated into the python cult, he obviously couldn't back up the treatment he was giving me with the proper sacrifices. According to *Nagan*, those sacrifices were indispensable for curing me mentally, while the herbs and roots would bring me physical relief.

'You're right, Gbalessu,' my father finally admitted.

He decided to take me the next day to Bè, to the priestess of the snake cult.

Next day, a little before sunset, at the hour when the day begins to cool, they got me up. I was helped across the yard by our *Fofogan* and another brother.

Father, wearing straw sandals and clad in his great woven loincloth, which he wore like a plaid over his shirt, walked in front. My mother didn't come with us. For the last few months she had been staying in her native village because she was expecting a baby; fortunately for her, as yet she knew nothing of my accident. There was no telling how she might react to the news of my condition, for each of the two daughters she had borne had died of fever, one before and one after my birth, at the ages of eight and twelve, so that I was her only child left. Moreover our mothers, who were considered only as child-bearers, had no place in their husband's dwelling (in which they hold no authority, being nearly always under the thumb of our aunts, our father's sisters), except when a living child formed a link between them and our father's family. On the other hand, they in turn had great authority in their brothers' households over their nephews and nieces. While my mother was absent from Lomé, my father's first wife acted as my mother. *Nagan* came weeping after us, together with one of our sisters who was trying to comfort her.

Outside, between our courtyard wall and that of the neighbouring patio, ran a narrow alley that today no longer exists; a few steps further it gave onto a section of dusty unfinished road, on which work had stopped several years before in our neighbourhood. It was the future 'Bè Road', which was later to link us with the mysterious village of that name.

We followed this road eastward to the point where it suddenly petered out. Crossing its final few feet, overgrown with wild grasses and weeds, we continued straight on, plodding through the sand and a broad wasteland which we used as an open-air latrine. This untilled ground seemed to have been enriched by the excellent fertilizer we contributed: it was covered with thick bushes from which rose giant flowering cacti, flourishing sisal plants, all kinds of thorny growths, and magnificent creepers carpeting the ground. After about an hour's walk across this terrain, in those days uninhabited, we came in sight of what is still known as the forest of Bè, which is really just the sad remains of the equatorial forest that once covered the whole of southern Togo, saved by some fluke from the deplorable deforestation our country has suffered. The peaceful village of Bè huddles at the foot of the clayey plateau of Tokoin, between the southern forest and a lagoon of stagnant water to the north.

For several generations a handful of animists, hostile to any new ideas, have lived in the woods. In sanctuaries protected by the shade of the virgin forest, they worship the forces of nature and jealously uphold tradition. Here is the gist of what our people relate about the past of these men with their fierce customs, and about the foundation of their village. A hunter named Djitri founded Lomé. His ancestors were caught up in one of the migrations which, radiating from the heart of Togo, brought them to the south. Djitri settled in a place he called 'Alomé', named after the trees which grew on the spot where his first hut was built, and whose branches provide *alo*, or toothpicks. Later the name lost its initial 'A' and became Lomé. In settling there, our hunter hoped to be safe from the wild animals which infested the entire region.

Some years later, about two kilometres to the east of Alomé, which was still covered by equatorial forest, Djitri founded for his eldest son Aglê a village which he named Adélanto, the hunters' quarter, and which later became Bè – and this is how. Some Aja tribesmen, emigrating from Dahomey (now called Benin) because of war, sought refuge in Aglê's new village. He obtained his father's permission to take in the refugees but they, fearing that their new home might be discovered by their enemies in Dahomey, passed a strange law: they would never speak loudly, never shoot rifles, and never amuse themselves by dancing to the rhythms of the tom-tom. For this reason Aglê called his village Bè, or hiding place. It was also called Badépé or Badékpa, 'enclosure of the quiet voices'.*

We shall come back to the Dahomean influence on the customs of these people. All we need say for now is that the inhabitants of Bè have always rebuffed the Christian missionaries attempting to evangelize them; that more recently they have opposed the continuation of the road to Bè – work was suspended in our own quarter of Kpéhénou – because it would pass through their sacred wood; that they were even against the introduction of electricity in their locality (their section of the town was always plunged in total darkness); and, finally, that they are the spiritual masters of Lomé. Because of the Aja's fierce determination, this sacred wood, the sanctuary of the

* Rev. Père Kwakume, *Précis d'histoire* (Lomé: Imprimerie de l'École Professionnelle, 1948).

most ancient altars consecrated to the region's divinities, still insolently flourishes, in all the original purity of tradition, only a rifle shot from the capital that still lies beneath its shadow.

Entering the wood was forbidden to members of other ethnic groups, except for urgent reasons such as consultations, initiations, or sacrifices. We knew that guards – the most ruthless of whom were often not the men but the women – were camouflaged among the trees to watch the paths. If an intruder ventured too far into the forest fringe, piercing cries would at once break out and echo through the trees with a kind of protracted 'ooh-ooh-ooh'. As they uttered this warning ululation, the invisible watchers would tap their open mouths rapidly at irregular intervals with the inside of their hands, fingers joined, in an urgent rhythm, giving the impression that these deafening howls were coming from several directions at once. This was to scare the intruder into beating a retreat, but also to warn the warriors to arm themselves if he continued to advance.

As for us children, in order to keep us away from the forest, we were told that any nosy boy or girl who ventured in there would meet Aguê, also called Azizan. This is the name of a fabulous creature of the bush who has only one eye in the middle of its forehead and only one arm; it also has only one leg, on which, we are warned, it can hop around with the greatest of ease and speed, ceaselessly patrolling all the forest paths. Its foot is back to front – that is, with the heel turned forwards, the toes backwards – so that its footprints deceive. Whenever it meets an intruder it has only to look him straight in the eye to scramble his memory. Then the intruder can't find his way back and wanders in circles until the medicine men come for him. We were also warned that as soon as you catch sight of Aguê you must take all your clothes off and begin to dance, before it can fix you with its eye. Apparently it entertains the creature to watch you dance naked; it doubles up with laughter and forgets all about you . . . This is the only way to escape its clutches. True or not, we knew that there were people who had entered that sacred forest and never been seen again.

Yet the morbid fear we felt on the outskirts of the forest didn't stem only from some possible meeting with Aguê the Sinister – whose name is curiously similar to that of Aglê, son of the hunter Djitri. Sometimes quick, harsh groans and plaintive lamentations came from the heart of the forest. These, combined with the awe the place inspired in us and the shrieks of certain birds that filled the wood, succeeded in turning our fear into indescribable terror.

It was nightfall when we reached that uncanny village. One last ray of the setting sun still glowed on the treetops. My father left us on the edge of the wood, giving us strict instructions not to budge an inch from where we stood. He took a gloomy path and disappeared. I was afraid for him and for all of us.

After a wait that seemed endless (black night was all around us now), he returned, carrying his sandals in his hand. He was accompanied by a tall girl about thirteen years old. She carried an earthenware lamp in which a wick of twisted cotton burned in a small pool of palm oil, its upturned end slightly overhanging one of the four beak-shaped corners of the lamp. The girl halted at the edge of the wood and refused to go any further. Later I found out that she was an initiate of the cult of Hêviesso, and that she had entered that sisterhood when she was barely eight years old. A stained white loincloth covered her from hips to ankles, leaving the rest of her body naked; she glistened with palm oil from her waist to her forehead, from the gentle slope of her shoulders to the tips of her fine fingers. Her hair, unbraided, was a great dark shock that seemed never to have known the touch of a comb, as with all the *vaudoussi* still serving a divinity. Two long necklaces of cowrie shells were twisted over her gleaming chest, which reflected the flickering flame of the lamp that fell on her prematurely serious child's face and the graceful contours of the most beautiful breasts I had ever seen: firm breasts with provocative upturned nipples. The girl appeared unaffected by the pitiful state she saw me in. Her broad forehead betrayed stubborn concentration, and her eyes were clouded by higher concerns.

Before conducting us into the wood, she said: 'You may not enter wearing anything on your feet.'

Nagan took off her sandals. We children had no shoes.

'No wristwatch either!' the girl commanded.

Father had already taken his off and put it in his pocket. I and my brothers had none.

The girl cast a final scanning glance to make sure we weren't about to enter the sacred forest with forbidden objects – especially imported articles manufactured by whites.

'Follow me,' she said at last.

Then, to the dull clink of the cowrie shells dancing on her breasts and at her wrists and ankles, she guided us into the wood through gloomy labyrinths that the lamp's dim glow could not penetrate. They told me much later that guards concealed in the trees and merging with the darkness had kept watch on our night procession.

We reached a clearing with a dozen low huts with mud walls and roofs of straw. As we approached the first of these, the girl suddenly uttered in a terrifying voice a piercing scream fit to raise the dead. That scream had the same effect on me as a cold shower. Quickly putting down the lamp, which nearly went out, she fell on her knees, bowed very low, then threw herself full-length in the dust, arms outspread, and lay motionless. Before we had recovered from our fright, seven men and women appeared from nowhere and formed a silent line behind us. Two young men slowly left one of the huts in front of us and moved forward like shadows to raise the girl from the ground.

They dragged her away, her superb breasts heaving as she panted for breath.

It was the last I saw of that beautiful priestess of the lightning.

A very tall man came towards us. As he walked he leaned on a long wrought-iron pike: at head height it had a kind of circular dish ornamented around the rim with half a dozen little oval balls, also of metal, and hollow. These balls contained objects which created a strange tinkling sound when the pike was shaken.

When the man was three paces away from us, he asked: 'Do you bring good news?'

'My son here . . .'

'Ah,' he interrupted my father, 'so he's the one.'

At a wave of his hand, the men and women standing behind us vanished into the darkness. Then this man who seemed invested with such authority took another step towards us. My father told him of my accident, from first to last.

After hearing out my father, the tall man began to speak. Upright and impassive, he punctuated each of his words with his pike, lifting it and bringing it down with a heavy thud, as if he needed that sinister noise to express his meaning.

'You will be led to the priestess of the snake cult,' he intoned. 'The road is not long. As you go, take care not to tread on the pythons during their peaceful evening airing.'

At these words, I felt I was going through a dreadful nightmare, or was already dead and transported to the regions of the evil spirits. I wanted to tell my father, 'Let's go home – I don't feel ill any more!' But this would have created a tricky situation with the forest dwellers and ruined everything.

Though he carried no lamp, the man was already moving down a path and beckoning us to follow. We walked behind him in single file. As if obeying the same reflex, each of us advanced with infinite care, treading in the steps of the one in front.

The hut where he led us was rectangular, and longer than the others we had seen. Its roof came three quarters of the way down the walls so that, without lowering your head, you couldn't see anything of the interior, which was only very dimly lit.

Our guide stopped on the threshold and said something. From inside came an almost inaudible reply. After some brief exchanges, we were invited to enter. Our guide then left us. My father moved forward, holding me by the hand. *Nagan* and the others followed us. We went in.

Over our heads hung three stuffed owls, tied by their feet to a roofbeam, alongside bats with outspread wings. Three of the four whitewashed walls bristled with the heads of various species of antelope with weirdly jutting horns. On the left-hand wall were horns of bushbuck with a single twist in them, fluted, almost straight; of reedbuck, curved forward in a quarter-circle; and of oribi, slender, black, sticking like a spike out of the forehead. Shells of giant tortoises leaned against the foot of this wall like shields. Above them there were rows of jackal heads; the bulky, bulging muzzles of warthogs with all their thirty-four teeth; crocodile jaws; and also bushpig skulls with elongated snouts, missing some of their forty-two teeth.

On the right-hand wall, beside rows of spotted hyena skins pierced with porcupine quills and leopard claws, you could see the hanging tails of horses and squirrels, then more horns: those of the addra or dama gazelle, short, fairly massive, swept back and finely fluted; of the damalisk or topi, their double curve set on a bony 'chignon'; and of the bubal, V-shaped.

The four pillars that supported the roof's central beam – where joists stuck out like the bones of a gigantic fish – were hung with monitor lizards, dried chameleons, egrets' and pink flamingos' heads, and pelicans with their long necks and pendulous beaks, everything covered with wild boar hides. There was a vulture, with its hooked beak and bald head and neck, and a big bird with dark plumage streaked with white.

The opposite wall, pierced by large holes through which the wind blustered, was loaded with the lyre-shaped horns of waterbuck, U-shaped red bubal horns, thick at the base and double-twisted, and finally the horns of roan antelopes, ringed, bent backwards and set in massive black and white striped heads. These latter horns alternated with buffalo skulls. There were also some amulets and various other things I couldn't make out at first glance. No occultist's hall all hung with black could have had so strange an impact as when we crossed the threshold of that murky den. A lamp, hung high up, cast a mournful glimmer over all these ritual objects and the low, sloping roof that had no ceiling; the dim glow plunged the beaten earth floor, all bumps and hollows, into shadows that thickened towards the corners. A strong smell of burning palm oil, mingled with the scent of herbs and other less definable odours, impregnated the whole place.

At the centre of all this, between the two central pillars, a thin, wizened old woman was seated on a mat on the ground, her legs stretched out in front of her. She stared hard at us for some moments, then rose to greet us and motioned us to sit along the wall where the door was, on a tree trunk stripped of bark and set on very low forked branches embedded in the ground. Our hostess used a calabash to take water from a big jar sunk up to its middle in the floor near the door. She drank a little of that water before offering us the calabash almost full. When this had made the rounds of the six of us, the woman took it back, took a couple of paces towards the door, and threw the rest outside, saying '*Ago-o-o-o*' in apology to the divinities she was afraid of splashing in the darkness outside. Then she came back and covered the jar. Finally she greeted us – until then we hadn't opened our mouths – and welcomed us again.

She resumed her position on the mat in front of us and started talking to my father.

It took us a good quarter of an hour to notice that in an unlighted corner of the hut, over to our right, there was someone who had been examining us intently for a long time. It was a big woman with huge dark eyes, squatting in the shadow, motionless as a statue. Father was the first to become aware of this strange presence: he suddenly stopped speaking. We all looked in the direction of his fixed stare, and there was a long silence. A dense, frightening silence. The woman did not say a word. However, everyone soon realized that it was upon me that she was focusing her attention. Indeed, I could feel the weight of a terrible unblinking gaze, like that of a person in a trance. When this woman leaned forward to examine me more closely, her head emerged from the shadows surrounding her – shadows that further darkened our long silhouettes against the wall – and I was able to examine her face at leisure: it was incised with ten ritual scars – two in the centre of the forehead, two on each temple, and two on each cheek – exactly the same number of signs that the initiates of the snake cult distinguish in the five creases on either side of the upper lip of the royal python they worship . . . The face of this corpulent figure would have looked rough and commonplace, had her eyes not glittered with a weird brilliance. On all of us she exercised a powerful, inexplicable fascination.

'Who is your mother?' she finally asked me calmly, without warning, not moving from her place.

'She is here,' I answered, turning and indicating *Nagan*.

She turned towards her.

'Is this your child?'

'Yes,' said *Nagan*, 'he is the child of my husband and one of my fellow wives.'

That big woman with the magnetic eyes had an astonishingly gentle voice, calm and steady.

'Was this the first time you had come face to face with a snake up a tree?' she asked, fixing me again with her eyes.

'Yes, it was.'

'My child,' she sighed, 'even in desperation to protect her little ones, the snake would never have attacked you as your father has just described unless, at some time in your life, you had done some wrong to reptiles.'

Here I translate by the word 'reptiles' a word she used that in Mina means literally the 'beloved children of the Earth'. We often use this term to describe reptiles who, unlike us, do not need feet in order to move on this Earth, the Mother of us all, whom they still touch with the whole of their bodies.

At the words 'wrong to reptiles' I could not help thinking of the poor lizards I had massacred in the past, who might now have been avenged through a snake. I was about to confess to this crime when *Fofogan*, guessing what I was going to say, snapped his fingers to attract my attention and signalled me not to speak.

'Yes,' that terrible woman went on, after casting a piercing look at my elder brother, 'perhaps a forgotten wrong, one that should have been put right at once.'

'Our Mother,' my stunned father naively replied, 'what possible harm could my son have caused to creatures we venerate?'

'That is for him to tell us!'

She turned towards me.

'Tell us how you've treated pythons whenever you've met them.'

As she spoke those words, her voice became strong and her eyes flashed. Here was the head priestess of the snake cult in person.

Realizing that it only had to do with pythons and not with other reptiles, *Fofogan*, who had been holding his breath, inhaled freely again, and his chest relaxed in a sigh of relief.

'I remember,' I told the priestess, 'that a few years ago, before our area was cleared of bush, pythons used to enter our yard freely. Often in the morning when we woke up we'd see one coiled around the central beam of the shed we use as a kitchen; sometimes he'd be hanging by his tail, head downward.'

'Then what did you do?'

'We used to chase it away.'

'How did you do that?'

'We'd imitate pigs grunting, because we thought pythons were afraid of them.'

'And the python moved away?' asked the priestess in surprise.

'Not always. So then we'd take a heavy stick . . .'

'To kill it?'

'Oh, no! We held the stick up to the beam where he was coiled and said to him: 'The pig's coming!' After a few moments' hesitation he'd slowly coil himself around the end of the stick until it bent beneath his weight. Then we'd tell him: 'Grandfather, don't ever come back to this house, because you frighten your grandchildren too much.' And then, singing a song, we'd go and put him back far off in the bush. But that didn't keep the pythons from returning to the house.'

'So you've never done them any harm?'

'Never!'

'But why were you afraid of them?'

'I don't know. I just . . .'

'Have you ever seen a dead one in the bush?'

'Yes, I've seen a dead one, but not in the bush. It was in town, in front of a European hotel. It had probably been killed out in the bush and was being skinned in front of the hotel.'

'*Ao!*' The priestess said this one word, hitting her chest violently.

She let out this groan with the sort of dreadful voice that

only women can produce to express pain when they hear distressing news. It was as if the priestess had felt the sharp blade of the skinning knife in her own flesh. What I had just described had happened a few years earlier, and it cast a terrible chill over my father and *Nagan*.

To understand their reaction, and especially the priestess's, it's necessary to have some understanding of what the python means to us. In fact, though it has the privilege, like many other reptiles, of being a 'beloved son of the Earth', the python, unlike other snakes, possesses many other striking attributes. He links heaven and earth: the golden patches scattered over his black skin recall the stars that sparkle in the sky at night. He is the image on earth of the rainbow that hangs in the air during a shower of rain. His movements resemble the flow of watercourses. Finally the python, symbol of the watery element and therefore of fertility, is a god. A Mina will never kill one, never eat one. On the other hand, we hunt and kill all other snakes, but we never eat them either. However, unworthy races all around us consider the python a great delicacy, and these we shun. It would be true to say that, aside from all political disagreements, the persistent hostility that divides our country's tribes is often caused by food taboos. Though it may not appear so on the surface, our relations are often governed by the saying, 'Tell me what you eat and I'll tell you who you are.' Brother or enemy – no compromise is possible. When a Mina sees a dead python in the bush, he buries it religiously, then makes for the sacred forest to purify himself by taking a bath specially prepared by the priests of the snake cult, for any man who sees a dead python or who kills one, even by accident, becomes unclean in our eyes and must purify himself in the ritual bath.

The python's true name, *dangbé* (from *dan*, snake), is not always spoken aloud. We often substitute the word *togbé*, grandfather. We don't know how this snake became a god to us, but tentatively it can be said that Dahomey, the great centre of the snake cult, is also its source. For the name Dahomey comes from *Dan Homé*, words which in Mina, our common tongue, mean – according to the inflection we give them – either 'the hut' or 'the belly' of *Dan*, the snake. So it is likely that the Aja, the migrants from Dahomey who sought refuge in the village that became known as Bè, arrived in their 'hiding place' bringing the altars of their household gods, which included the python. These new gods easily found a place among the already numerous ones existing in the region, and eventually the python became a god for us, too.

So there was a great silence when the priestess, hearing that I had seen a python skinned outside a European hotel, uttered that cry of suffering. Then she said to me:

'Don't you realize that the sight has defiled you? Poor child, you must be purified!'

They made me take off my loincloth. The skinny old woman who had greeted us took me by the hand to lead me outside for the ceremony of purification. At that instant, I saw something unexpected and flung myself back with a cry of alarm. One of the skins (I don't know of which animal) hanging on the wall opposite suddenly began to shake, as if blown by the wind, then slowly came away from the wall to reveal a moving body. It was a huge python entering by one of the numerous holes made in the wall. His long body slid on and on from behind the animal skin, like a sword coming out of its sheath, and all the while he slowly poked his flat head from side to side, seeking a way through the mass of objects hanging on the wall. We stared at him, transfixed. He let his neck glide over one of the antelope horns, at the same time hauling himself forward. When his long body had completely emerged from behind the skin, he let himself hang from the horns, like an outsize cable broken in the middle, with its ends trailing on the ground. The priestess watched him with ineffable delight.

'Go and see if they have all come back,' she told her assistant.

The woman left me standing naked in the centre of the hut and hurried to the door. She had just crossed the threshold when another python entered, slithered right across the room with a shuffling noise as of a man dragging his feet along the ground, and at once disappeared behind the tortoise shells, leaving us with the feeling that we had been dreaming. But a door creaking shut outside told us it was no dream. The assistant returned and told the head priestess: 'Yes, they've all come back.' Then she asked me to follow her.

We went out into the wood. Bright moonbeams flooded the clearing. We walked barefoot through the grass for about forty metres, then stopped outside a small circular fence made of wicker hurdles and branches attached to posts. At the open entrance to this enclosure, the woman prostrated herself and made a sign to me to do likewise. Then, without wiping the sand from our foreheads, we went in.

Inside, there was only a *canari*, painted with vertical red, black and white stripes and held in the triple fork of a leafy tree planted in the middle of the enclosure. The *canari* stood about one and a half metres off the ground and was brimful of water, which gave off a strong smell of macerated plants. In it I recognized some broad *kpatima* leaves and numerous roots.

'You will wash yourself with this water,' the woman told me. 'It contains plants that will do you good. You will not dry your body after this bath: the water contains other plants, and breathing their aroma will calm you for the ceremony to follow, which will put you on good terms again with those you have offended. This earthenware jar must not touch the ground. No metal object must touch this water. Come here.'

A bowl was floating on the water. She used it to pour water over my head, saying:

'You are a child, the fault is not on you. It is on the whites, and shall evermore fall upon their heads! What! Is nothing sacred to you – you of the white skin and hair of straw? Skinning Grandfather in front of this little boy not yet even hatched from his groundnut shell!'

The water was very cold. I thought it inhuman to use it, in the middle of the night, to bathe a feverish child.

After the bath, we returned to the hut.

I was cold: I was shaking and my teeth were chattering. I was as naked as an earthworm, the cold water still dripping from my body. Pieces of green leaf clung to my hair.

During our absence, the priestess had moved. She was now squatting in the middle of the hut, on a little bench placed in the centre of a white circle traced on the ground with kaolin. Her white loincloth (which only the *vaudoussi* wear) was knotted above her breasts. When we came in she was talking to my father.

My companion took the only lamp in the hut and gave it to me, with a request to follow her outside again. We went behind the hut and approached a structure made of beaten earth, with a flat roof. Cautiously, the old woman opened the door and beckoned to me to bring the lamp closer. It was so dark inside that for some minutes I could make out nothing. The wind was blowing, and the lamp shed only a faint flickering light all around us. When it finally lit the place enough for me to see what it contained, I could hardly believe my eyes. Coiled up one on top of the other and interwoven like a mountain of spaghetti, a great cluster of pythons, slithering under and over one another, were slowly digesting heavy meals beside large pools of palm oil, blinking their cold eyes and darting their forked tongues. The woman picked one up, an enormous one, and the snake at once wound its gleaming black and yellow speckled coils round her arm, letting the rest of its body trail on the ground. The woman lifted it up and placed it around her shoulders, then we returned to the hut. I kept my distance from that weird woman. What must I go through next? I'd had a bellyful.

As soon as she saw us on the threshold, the priestess exclaimed, as if in a trance:

'*Mia wé zonlo!*' (You are welcome here!)

'*Yoooo lo!*' her acolyte replied.

Suddenly rising to her feet, the priestess came towards us without leaving the white circle. Arms outstretched in a clinking of cowrie bracelets, she received the python; then she sat down again on the stool, settling the snake on her knees. The lamp was put back in its place. The horns at once resumed their strange appearance in the shifting shadows, ghostly and threatening. Inside the circle, *djassi* – maize flour mixed with water – had been poured on the ground: a mixture used in many ceremonies.

The python stayed only a moment in his mistress's lap, then uncoiled, slid slowly to the ground, and wound himself around the pool of milky-white water, where he remained, lethargic and unmoving.

'My child,' the priestess said, 'come and kneel down here.'

I knelt only two paces from the snake, whose little black eyes, fixed and glittering, gleamed with the same light as his mistress's. My heart began to pound.

'Come closer and bow your head.'

I bent my head, closing my eyes in fear. At first I felt the touch of a cold liquid. Seven drops of it fell on the nape of my neck. Then a heavy, icy body settled heavily on my shoulders.

'Stand up, child.'

Eyes still tightly shut, fists clenched, I slowly stood up. My shoulders bent beneath the weight.

'Don't be so stiff, so rigid. Unclench your fists, relax your jaw.'

I just stood there, trying to keep as calm as possible.

Like a long, thick liana, the python wrapped itself round my right leg and then, sliding his head slowly over my collarbone, he glided like a trickle of cold water down my spine. With each extremely slow and spaced-out breath, he seemed to be tightening his hold around my leg. Sweat beaded my forehead. I felt outraged, because my father was there and yet did nothing to spare me this horror. I could understand not a word of what the priestess told the python in a vehement and emphatic monologue spoken in a secret tongue. At one point the reptile, twisting his tail from out of his coils, gently slapped my right calf with it as if to express the pleasure he felt at the contact with my warm, naked body. His forked, slimy tongue licked around my neck, in my ear, and over my face. This god of my ancestors betrayed no sign of aggression. Finally, as the flow of the priestess's words died down, he dropped to the floor.

The wrong had now been washed away, they told me. I sincerely believed so.

It was far into the night when the ceremony ended. They told us that it was now too late to leave the sacred forest, and spread mats on the floor so we could spend the rest of the night in the hut. Before we went to bed, the priestess gave me some advice about my future attitude to snakes.

'People who don't understand the first thing about life,' she said, 'chase them on sight. They wickedly wound and kill them. But these beasts are both the gentlest and the most unhappy creatures on this earth. Look! I always go to bed with mine, and they have never done me the slightest harm. A snake bites only when trodden on or disturbed. In that respect, they are no different from us. What happens if you accidentally step on a man's foot? He nearly always turns on you, ready to slap your face; only by apologizing can you calm him down, and even then . . . Can you imagine what men might do to one another, if they had poison fangs? Yet just think, no one ever apologizes to a snake – let alone expresses any kind of regret for all the sufferings inflicted on them. Do people even realize what harm they do to snakes when every year they keep destroying a part of the bush that is their natural habitat, and chase them away from round their houses? The poor things are so surrounded by enemies, they spend their whole lives trying to escape. For besides man they also have to protect themselves from attacks by enemies no less to be feared, the ants.'

'Ants?'

'Yes, ants! And even a new-grown shoot of grass can hurt them! Grandfather suffocates his prey in his powerful coils, breaking the ribs. He swallows it slowly, swallows it whole, after coating it with saliva. Then he has to keep absolutely still for a time, because it's hard for him to move while digesting his prey. This enforced stillness may last several hours, sometimes whole days, according to the size of the animal swallowed. Even the body of only a medium-sized animal, swallowed bones and all, prevents him from moving freely at once. Sometimes a trickle of blood comes from the mouth or the nose of the victim as the python is squeezing it to break the ribs; then, while he's lying inert to digest it, columns of ants, attracted by the blood, will fasten on the python. They enter his mouth, his nose, swarming all over the softer parts of his flesh, stinging him all over. He dies after the most terrible agonies. But the python, who is very clever, doesn't always let himself be taken so easily. He can go without food for a very long time, and so reduces these risks. Often he even leaves his victim half dead, then crawls in a wide circle round the wood to make sure that certain kinds of anthills aren't there. Nor will he swallow large prey in a freshly harvested stubble field. The young shoots, which are very sharp, may start to grow again and pierce his body, if he has to lie there helpless for too long!'

I don't know if all this was true or if the priestess was trying to make me feel sorry for pythons. Be that as it may, my eyes were now shining, not with fear but with compassion.

'See what they endure in this world! If you find one lying dead in the bush, dig a hole. Then gather leaves, lay them in the grave to make a bed for him, and bury him. Afterwards, don't forget to come here and wash yourself, purify yourself. The wrong you righted tonight, which made you impure in the sight of the gods and the snakes, was committed by others. But it still made you unclean.'

She saved the most important part of her speech until the end. Before rising from her stool, she told me:

'I believe that you will never harm the beloved children of the Earth, especially the Grandfathers, because, from what I've observed, you are sensitive to the sufferings of others. With such a frame of mind I'd say that, for your own sake and to carry on the ancient tradition, you ought to take to the bush for a few years to become a priest of our cult. Only, it's a shame you're over seven years old, because you should enter our order at a time when you're still innocent in every respect. I myself was promised to the python when my mother was still carrying me in her womb. All the time she was pregnant with me, my mother was besieged by snakes. They gave her no peace until they were solemnly assured that the child about to be born would be dedicated to them. Still, if you're sensitive enough, you can recover all or part of your lost innocence. Come back and see me as soon as you are better.'

I had always thought that the worship of the python was

something reserved for the Pédas, a Mina sub-tribe, whose members are not necessarily priests of the cult, but who from childhood on, without distinction of sex, bear the ten ritual scars that marked the face of the priestess. I didn't know that the other sub-tribes of the Mina could be raised to the priest-hood. So the final words of the priestess took me aback. It was not for me to make the decision. If my father ordered me to do what she asked, it was my duty as his son to obey him. I turned to him to hear what he would say. For the moment he was silent, but the priestess repeated:

'Come back and see me as soon as you are better.'

Then she stood up and returned to her corner. Her assistant had long since taken the python back to its little building, and had just unrolled two more mats.

The two women spent the rest of the night in the shadows, while we slept in the centre of the hut. Exhausted by the night's emotions, I fell into a deep sleep.

Next day, we were up with the sun. They gave us *akassa* to eat, then the priestess gave my father a list of plants for my continued treatment. So it was that, to express his gratitude, he promised to bring me back to the sacred forest as soon as I showed signs of recovery, to have my head shaved so that I could be accepted into the bush as someone 'newly born'. At the same time he would thank the priestess for my cure, which he would repay in kind, meaning sheep or chickens. For one does not actually pay the healer until the sick person has recovered his health.

'The gods are the only healers,' replied the priestess in a detached tone of voice.

'Very well,' my father answered, 'I shall come and give them my thanks!'

Shortly afterwards we set off for home.

I stayed in bed for two weeks more, and all that time was

subjected to a rigid diet of bitter plants whose infusions had been recommended by the priestess. Little by little my strength returned; I was able to get up, and come and go in the courtyard on my own. People told me, 'You're almost as frisky as a kid again!' And they naturally attributed this return to health to the cure prescribed by the priestess. They were just waiting until I was fully recovered before taking me into the sacred forest again to put me in the hands of that woman, my future initiator. In other words, I was already pledged to the gods who had saved my life. Father collected a number of the white loincloths for me to wear in the bush once the cult members had shaved my head. To pay for my cure, he picked out from among his sheep, first, a villainous-looking ram with twisted horns still intact despite constant bitter battles against other males in the yard for possession of the females; then two mild-eyed ewes – all three with repulsive, dirty fleeces. It was a waste of breath for *Nagan* to remind him that people in our locality disapproved of offering someone an uneven number of gifts, and that, considering his reputation in the neighbourhood, and the happiness he must feel at having a child's life spared, particularly a son's, he ought to throw in a fourth sheep: father refused. I think it was out of thrift – we were twenty-six brothers and sisters in our family, not counting our mothers, Uncle Ahouanssou, and my father's three servants. But he said to *Nagan*:

'That woman is a priestess. The uneven number, which is a bad omen for gifts to ordinary people, is acceptable to divinities, so it must also be acceptable to those who serve them. I myself, for example, never take more than three chickens when I treat a sick person!'

He would not budge from that position.

Our return to the sacred forest was set for about six weeks hence. To shorten my convalescence, two of my elder brothers who, like myself, had no classes at that time, used to take me walking along the beach, which was about three hundred metres from our house. In our compound there was pride at the idea that soon our numbers would include a serving priest of one of the traditional cults. But an insignificant, unexpected event was to rob my family of that fond hope.

3

The Runaway

The Evangelical Bookshop, in the commercial quarter at the other end of town, was one of Lomé's modest bookstores, run by missionaries. As the missionaries concentrated on the conversion and instruction of an illiterate people who were mostly pagans and idol-worshippers (there are very few Muslims in the country), the shelves of their shop were laden with school textbooks and religious primers. However, from time to time, as if in error, the occasional travel book or a novel would find its way onto the shelves, and, every week before my accident, I had fallen into the habit of spending the money I made by selling my mats on books which were never on our school syllabus.

So one morning when my brothers had left early for the coconut plantation, and there was no one left at home to take me for a walk along the seashore, I went out alone and visited the Evangelical Bookshop. Inside there were two shelves against the walls on either side of the counter. I went up to one of these, attracted by a book laid flat on a half-empty shelf, with a cover showing a picture of a hunter dressed in clothes made of animal skins and leaning on a spear. I was struck at once by the title: *The Eskimos from Greenland to Alaska* by Dr Robert Gessain. The book was illustrated with photographs and engravings: I liked the look of it, bought it, then went on my way to the beach. By noon I had finished my new book, the first I had read about the life of the little men of the North.

Was it the author's praise of their hospitality that triggered my longing for adventure, or was it fear of returning to the sacred forest? I hardly remember. But when I had finished reading, one word began to resonate inside me until it filled my whole being. That sound, that word, was *Greenland*. In that land of ice, at least, there would be no snakes!

'All Eskimos,' the author wrote, 'live in an Arctic climate essentially characterized by the alternation of two very contrasting seasons. The winter is very long, excessively cold and dark, and the polar night is its most striking feature . . . The Arctic climate is defined by a temperature which does not exceed a mean of ten degrees during the warmest month.'

I was unable to imagine such a temperature, and began dreaming of eternal cold.

Lying on the hot sand, lapped in the torrid humidity of a tropical noon, the contrast with Greenland seemed to me all the more remarkable.

'Within the geographical limits of the Eskimo habitat no trees grow, except for a few dwarf willows that emerge only to creep along the ground.'

No trees! And real men had been living in that land for thousands of years?

The illustrations showed these men plump and smiling. Strange clothes made of animal skins covered their bodies, and all you could see under their big hoods fringed with thick animal fur were their happy, open, honest faces. As the Eskimos were hunters, I had no doubt that, thanks to my own hunting experience, I could make a living among them.

Underneath a photograph of a child putting something in his mouth, I read: 'Eskimo boy eating raw fish.' Poor thing! I thought, suddenly thinking of our own appetizing dishes, always highly spiced. But I also learned that in this far-off land the child is king, free from all traditional and family restraint – that was worth more than our well-cooked dishes. However, my stomach was accustomed to food that was not raw but carefully cooked. Could I take Eskimo food without disastrous consequences for my constitution?

Before giving serious thought to this crucial question, my mind was made up.

But it was not enough just to get the idea of going to Greenland. It was important to find out how to get there, to find the country on the map and see its position in relation to Togo, my starting point. Next morning, I returned to the bookshop and asked for a map of the world. The one they showed me was in colours ranging from darkest green to brightest yellow, indicating the characteristic vegetation of each country: equatorial forest, temperate forest, monsoon forest, Mediterranean vegetation, steppe, prairie, wooded savanna. In the midst of this fine, warm range of shades, Greenland appeared with not a touch of inviting colour: only a greyish border, indicating the tundra, traced a grudging line round three quarters of the country's area; the whole interior was a uniform white. Another book told me its size: 2,175,600 square kilometres, the biggest island in the world. Lying to the north of America, Greenland extended in one enormous mass of ice to the Pole.

It seemed further away from Togo and Africa than I could ever have imagined. Never in all my life would I have enough money to travel there directly! And even supposing my father spared me from returning to the sacred forest, his modest salary as foreman at UNELCO (Union Électrique Coloniale), the Overseas Electric Union, could never pay for the journey. Anyway, he would never let me leave for Greenland at the age of sixteen. So, to carry out my plan, I could only act in secret: I would have to run away from home.

I had seen plenty of Hausas, who are a semi-nomadic people living mainly in Nigeria and Niger. They travelled all over West

Africa carrying a bag on the end of a stick laid over their shoulders, full of medicinal powders and roots, the stock of an age-old trade which they supplemented by occasional farming and rearing livestock. Resting during the sun's worst heat, they travelled mainly at night. If I were to proceed as they did, in easy stages, finding some sort of work for a year or two in each country I passed through, I was bound to reach the Eskimos one day! I would have to begin with Ghana, the country next to Togo, then gradually travel up the coast of West Africa, cross the whole of Europe, and sail from Denmark.

Then I remembered that my aunt, married to a Ghanaian fisherman, was living at Port-Bouët, a few kilometres from Abidjan. An aunt, on the father's side, personifies divinity itself for us: her curse upon a nephew or niece is one of the greatest misfortunes imaginable in my region, just as her blessing is the happiest of omens; and so strongly do we believe this that when the orders of father and aunt conflict, we often follow the aunt's. The father's numerous half-sisters do not have this power, only his true sisters. So we sometimes call these aunts 'our fathers', although they are women and although we already have a word to designate a paternal aunt, *Tâssi*. I was on excellent terms with this aunt in Abidjan, who was a true sister of my father. He couldn't hold it against me, when he learned of my departure, if I had gone to 'pay her a visit' in the Ivory Coast. I wrote to her and in the month of August, only a week before I was to return to the sacred forest, I left for Abidjan in a truck, without a word to anyone, not even to my mother. It was one of those trucks called 'bush taxis' – kept going by second-hand spare parts, and creaking horribly on its battered springs – that tackled, as best it could, the first kilometres of my journey towards the Arctic.

The small sum I had saved since the beginning of the holidays by selling mats was only enough for half the journey. To pay for the other half, the driver employed me along the way as his 'apprentice driver'. This curious job involved riding on the roof to keep an eye on the passengers' luggage. Among the sacks of flour and chicken cages tied in a jumble on the roof, I made a kind of nest to protect myself at night from the wind that whipped my face and turned suddenly cold at sunset, in violent contrast with the overpowering heat of the day.

There were some nasty surprises along the way – they were frequent on our antiquated public transport. One of the truck's wheels flew off when we were travelling at top speed; a door held on with wire came away when the continual jolting snapped the wire; then flat tires followed one after the other with alarming regularity, not to mention the leaking radiator. But in the end, after a hair-raising trip along a dusty, winding, potholed road, we arrived safe and sound, three days later, in Abidjan, the first stop on my long journey.

Welcomed warmly by my aunt Adjoavi, I lived with her at Port-Bouët before finding, one month later, a job at the UAT (Union Aéromaritime de Transport) agency in Abidjan. It was too far from my aunt's house to where I worked, and I had to live in town. So I moved into a room in the Treichville quarter, delighted to be living alone and independent for the first time in my life.

No sooner had I gained this freedom than my aunt tried to take it away. On her first visit to my new lodgings, she slapped a packet of smoked fish on my table, sat on one of the two chairs in my room, and ran a disapproving eye over the general mess.

'Now that you've found a job,' she told me, 'all you need is some nice young girl to take care of you, someone well brought up, like all the girls in our village. It's not good for a young man starting out in life to live alone; he gets carried away and does silly things. I do hope you won't be like that. Don't spend all your money. As for your food, I'll bring you fish and rice

regularly. Save all your money, so that in a few years' time you'll
be able to build a house back home and make a good marriage.
You've started work very young, so you'll find success before a
lot of others.'

'*Tâssi ammî*' (Blessed be thy words, Aunt), I replied, sticking
to the usual formula for that kind of advice.

She paused a moment, then said:

'The next time I go home on a holiday, you'll write a love letter without putting a name on, and give it to me in an envelope.
Once I'm back home I'll find the right girl for you. Then we'll
add her name to the letter, and someone in her family will read
it to her.'

This prospect frightened me so much that for the first time I
looked my aunt boldly in the eye and said:

'No girl I haven't chosen myself will do for me!'

'Don't worry,' she kept on, 'I'll pick you out a good one, my
son. Don't you trust me?'

'I trust you, Auntie. But right now I have other plans.'

At that, her eyes fell upon the little shelf hanging over my
table, where I kept the few books I owned.

'Oh, I see,' she went on. 'You want to be like the other youngsters, and leave your home and go to France?'

'And why not?' I asked, making a great effort to conceal my
true destination.

My aunt, of course, had never heard of Greenland or of
Eskimos. But if she found out that I intended to travel to a
country farther away than France, which to her vaguely represented the whole of Europe (in Mina the name for France is
Yovodé – from *yovo*, the whites, and *dé*, country – land of the
whites), she would refuse me her blessing. And I preferred not
to have that blessing than hear my aunt pronounce a curse
which would bring me trial and tribulation. She leaned back in
the chair, crossed her arms, and said:

'I hope you won't set out for France without leaving a replacement in the family, a child to remind us of you? Don't go off on this journey too hastily – you're still very young. Well, anyhow, we'll talk it over again,' she concluded, rising from her chair.

My course was clear. I would have to set off at the first chance on the second stage of my journey, before my aunt brought me back a home-grown fiancée!

My worries didn't last long, for one day my aunt came to warn me that there was trouble in store for foreigners working in the Ivory Coast and that it was quite likely we would soon be deported, along with the Dahomeans. It was a fact that too many immigrant Africans (mainly Dahomeans and Togolese) had worked themselves into key positions in the Ivory Coast at the expense of the native population, and people were saying that the only remedy was to send all the 'foreigners' home. Official notices confirming this news were posted on December 1, 1958, and two weeks later I landed at the airport in my native land on board one of the DC-4s used to transport the thousands of foreigners expelled from the Ivory Coast. The 'refugees', as we were called, were not allowed to choose any destination other than their homeland.

Back in Togo, I acquired what might be called a new rank in the family hierarchy, because I had worked abroad and paid my own way. Prematurely regarded as an adult, from now on I could do as I pleased. So three months later I left without hindrance for Ghana.

It was on the eve of our countries' independence in 1959. The indignation provoked by events in the Ivory Coast died down in the face of that great new political upheaval; a strange new wind of brotherhood was blowing through Africa. The idea of a future continental union, put forward by Ghana, an English-speaking enclave in a group of French-speaking territories, suddenly drew together peoples which had long been separated. Accra, the capital of Ghana, became the centre of attraction. Adventurers of all kinds flocked there.

In that city I first of all enrolled at a cultural centre in order to perfect my English, which would be useful on my journey, and at the same time I attended the Alliance Française to fill in the gaps in my general education. Soon I got a job at the new embassy of Guinea as a bilingual typist, and worked closely with Camara Laye, the first ambassador of Guinea to Ghana and the famous author of *L'Enfant noir* (The African Child). (I had learned to type because, in sharing out the domestic work at home, my father had saddled me – among other things – with writing his and the family's letters.) A little later, 'The Voice of Ghana' radio started a French service; one of the first recruits, I used to work there in the evenings, keeping my daytime job at the embassy. In that English-speaking country, the new regime of Kwame Nkrumah was appointing Africans who could speak both French and English to important positions, and many of my Togolese compatriots sought a future in those short-lived pan-African institutions. Beside them, I seemed quite ridiculous with such different ambitions.

In June 1961 I heard that a ship belonging to the Chargeurs Réunis line would be calling at the Ghanaian port of Takoradi in two months' time. The following month I handed in my resignation to both employers and, the night before the ship sailed, went by road to Takoradi and spent the night in the port. This time, my savings took me further: I paid my passage as far as Senegal.

On board, the food must have been excellent, for the head cook was French. But my stomach was so badly upset by my first sea voyage that I couldn't sample any of the meals served on the ship, complete with wine – all included in the price of the ticket.

All the same, when I arrived at Dakar, I felt a double satisfaction: first of all, because I had just come a big step closer to my goal, and secondly, because I had been in a hurry to leave Ghana, whose nearness to my own land constituted a serious threat to my plans. In fact, my new adult status in the family hierarchy obliged me, according to custom, to make effective contributions to our village ceremonies. So an uncle could drop in on me at any time – a poor prospect for my budget! But in my determination to escape these probable invasions, I had spent all my money on a journey that would take me as far from home as possible, making only brief calls at the Ivory Coast, Guinea, Liberia, Sierra Leone and Gambia. Then at Dakar, after taking a taxi from the harbour to a Lebanese-run hotel, one of the cheapest in town, I found myself in my room with my luggage and virtually no money in my pocket.

I used what little I had to buy food in the street – some bread, a spoonful of margarine and a hard-boiled egg. That afternoon, confined to my room for the rest of the day, I was worried stiff. How was I going to pay my hotel bill? I spent a sleepless night.

Of all the solutions I considered, one kept coming to mind. I had noticed the embassy of Ghana on the rue Félix-Faure. Perhaps my knowledge of English would come in handy.

At the embassy early next morning, a charming Togolese girl at the reception desk informed me that for the last few days there had been a position vacant as newspaper translator. 'Wait a moment,' she added, going towards a door. She came back and ushered me into the under-secretary's office. Satisfied with the results of the test he gave me, he took me to the first secretary. Both were Ewes, natives of the former British protectorate of Togoland, my compatriots at the time of the German colonization. They expressed great interest in my desire to get to Greenland and introduced me to the chargé d'affaires, to whom I once more told my story. In the end, I was offered the position of translator. To tide me over, they made me an immediate advance of fifteen thousand CFA francs. And that wasn't all: the embassy had rented a furnished house for the Ghanaian chauffeur of the chargé d'affaires; they thought it too big for him alone, and suggested that I should have two rooms.

I could have stayed a year or two in that embassy and saved enough money to go straight to Greenland, but I was afraid that this easy life might finally distract me from my goal. Six months later, I handed in my resignation and left for Mauritania.

From Nouakchott, the capital, I went by Land Rover to Port-Étienne (now called Nouadhibou), planning to cross the desert to Algeria. A disappointment awaited me at Port-Étienne: vehicles no longer went that way. I learned that travellers intending to cross the desert to Algiers took the Tamanrasset road, starting from Agadez, in Niger, and not the road from Mauritania. Niger is to the north of my native land: there could be no question of turning back so far. Since I couldn't cross the Sahara on foot, I had to give up that idea and find another solution – to take a ship to Europe. But Port-Étienne, a fishing port, was not on the steamship route; so after working in the Peyrissac factories in that town, I returned to Dakar, where I had to look for work again. This time I got a job at the Indian embassy.

The lack of transportation across frontiers and the urgent need to find new work at each stage of my journey were not the only difficulties I had to overcome: there were also my studies, which I continued by correspondence while travelling from place to place. During the first few years my lessons came regularly by post, and all was well. But because of my frequent moves, my exercises, corrected in Paris, were seriously delayed before reaching me. So I decided to teach myself, as this seemed to be the best answer for a wanderer like myself, and I embarked on a thorough study of all the French classics, beginning with the sixteenth century. My large suitcases eventually contained more books than clothes.

Sometimes I talked about my plans with young Africans my own age. Some of them thought I was absolutely crazy; others, that I was wasting precious time travelling and just throwing my money away. And they would add: 'But how much money will you make from this journey, once it's over?' As if the only reward could be in cash!

The most surprising opinion was one expressed by a marabout in Dakar. This old man told me:

'You must have been born in that land of ice in a former existence. That's why you're going there – to return to your origins.'

That made me smile, for it was hard for me to imagine myself as an Eskimo!

'I think it must be your ancestors' nomadic instinct that is reviving in you!' suggested Guy Echevrery, a French friend, and the only one until then to give me any real encouragement.

According to him, my stay with the Eskimos would be of great interest because of the comparisons I could draw between Eskimo customs and African traditions – comparisons that no one had previously made. Guy gave me some letters of introduction to use when I arrived in France.

Six months after my return to Dakar I quit my job at the Indian embassy, and one month later, on May 2, 1963, I finally embarked for Marseille.

And that is how, in this era of interplanetary flight, it took me six years to get out of West Africa.

4

First Steps in Europe

In the course of the voyage to Europe we stopped twice in North Africa, at Casablanca and Algiers. In those two cities any African passenger who just wanted to go ashore for a few hours was required to hand his passport to the police, and he didn't get it back until he re-embarked. Those who didn't have a passport were categorically forbidden to leave the ship, whose gangway was never left unguarded. Apparently, freedom of entry to other African countries had not been laid down in that ambitious programme for unity; it soon became obvious that it existed only in leaders' speeches! Ironically, we could move across frontiers more freely in colonial times. Now, because of an absurd nationalism springing up between supposedly brotherly neighbours, each country in Africa insisted on passports and visas, inventions of the whites, while denigrating these same whites to their people. Happy the old days when it was simply tattoos, ritual scars and language which gave our fathers free passage in an Africa without frontiers!

As an exception on this voyage there was a third port of call, Livorno, on the southern coast of Italy, which delayed us for a day. Finally, on the sixth day after leaving Dakar, the ship dropped anchor at Marseille. Already the air was much cooler, but it smelled good.

I landed merely by showing my identity card, and found that France is a hospitable nation: despite the storm of ill feeling at the time of our countries' independence, no restriction was imposed upon our entry into the former mother country. Because of this freedom from formalities enjoyed by the peoples of the old French colonies, I felt freer in France than on African soil. But the quantity of ships moored in that great port, the incredible number of cars and fine houses, and the busy lives of the inhabitants all combined to give the vivid impression of braving another civilization.

I stayed only one day in Marseille, enjoying for the first time the pleasures of the carefree tourist. After a pleasant night at the Hôtel de la Poste, in a fine central district, I took the early morning train for Paris: comfortably settled in a second-class compartment, I was able to admire the most varied landscape I had ever seen, on a journey that took seven hours.

The high ranges of the south, covered with plushy vegetation, gradually gave way to the beautiful central plains. In that month of May the entire countryside, adorned with a foliage more tenderly and translucently green than on our trees at home, resembled one vast garden. Throughout the journey, I saw no region that was absolutely dry or deserted, nothing to recall those great, arid, stony spaces of our continent in which flat rocks jut from the ground like so many thirsty tongues panting for a drop of rain. Early in the afternoon the train entered the suburbs of Paris and soon came to a halt in the Gare de Lyon.

A great swarm of passengers got off the train and flowed slowly towards the exit, literally sweeping me along with it. There, anyone who wanted a taxi had to stand chafing in an interminable line, awaiting his turn. I took my place in the line and got a taxi about half an hour after the train's arrival.

In Dakar my friend Guy had given me three letters and, as they were addressed to Frenchmen living in Paris, it never entered my head to look for a hotel. I picked one letter at random: it was addressed to Monsieur Claude Géraudel, a former chief administrator overseas. I didn't know that, contrary to the custom in my own country, you are supposed to telephone before calling on someone with your introduction. In fact, in my family, a stranger who comes on the suggestion of a mutual friend and who gives notice of his arrival in advance is received with great reserve; we feel that his attitude denotes a lack of trust unflattering to his hosts. There is always room in an African home: the stranger has only to accept our food, our sleeping mats – in other words, our way of life.

Thinking perhaps that I was still in Africa, I simply gave the taxi driver the address of Monsieur Claude Géraudel, who lived in the sixteenth *arrondissement*. Then I gave all my attention to the impressive spectacle of my first ride through Paris. The taxi left the immense greyish station, turned into a boulevard, and drove towards the Seine. All along the streets stood blocks of imposing stone buildings that rose higher than the treetops and stretched into the distance like endless walls. When I raised my eyes, I could see only thin strips of faintly luminous sky against which the silhouettes of roofs laden with a host of pink sandstone chimneys and television aerials stood out. Lines of elegant wrought-iron balustrades separated the numerous storeys and ran beneath the twin-shuttered windows. Only these shutters were open; behind the windowpanes hung the opaque folds of white curtains. There were no idlers leaning on the railings. Sometimes clear sunlight would strike the tops of buildings, leaving the bases in shadow; at other times the whole street was filled with light. The broad, straight avenues offered charming distant perspectives often leading to some majestic monument. The bodywork of cars a long way off shone in the clear air like golden sequins caught in a torrent roaring through deep ravines. The constant noise was quite different from the light rustling of our palm leaves: the dull rumble of thousands of vehicles travelling bumper to bumper and giving off a strong stink of exhaust fumes.

The taxi crossed the Pont d'Austerlitz and joined one of the four lanes of vehicles moving along the quays. Now we were stopping for a red light. On both sides of the street, a busy crowd walked up and down the pavements. Others were crossing the street in both directions, some of them almost running. The men all wore grey suits, without any bright colours. They were walking fast and seemed on edge. The women had their hair dyed all kinds of colours. In the general hustle and bustle, they walked along with their shoulders hunched forward, looking up only to flash a stealthy sideways glance from time to time. I couldn't see any with the supple, noble walk or majestic bearing of our African women. Doubtless the pressure of Parisian life forced them into that nervous, jerky way of walking.

What intrigued me most of all was precisely what I couldn't see: the Métro. Was it possible that at this very moment a multitude of human beings were walking, riding and above all breathing under the ground, perhaps even under the wheels of these cars? I had the impression that I was living in two cities, one on top of the other.

Turning towards the other bank, the taxi crossed another bridge, from which I glimpsed the Eiffel Tower. Then, going down an avenue, it soon dropped me at the address of Monsieur Claude Géraudel.

Left alone on the pavement, I couldn't get the front door of the building to open. After a couple of good pushes I waited in front of it for more than a quarter of an hour, from time to time giving it a discreet little shove. A group of youths happened to go by while I was trying to push the door open; when they had gone past, they turned around and burst out laughing. A few steps down the pavement was an optician's shop whose owner kept popping out onto the doorstep. In the end, surprised to see me still standing there by the door, with my suitcases lined up along the wall, he asked me:

'Have you tried pressing the button?'

And indeed, on the wall to the right of the door was a shiny copper plate with a button. I pressed it: there was a click and the heavy double door of carved and varnished wood opened of its own accord. 'Pretty sharp, these whites!' I said to myself as I went in.

I found myself in a hall whose inside door was also closed. There was a marble plaque fixed to the wall, with a dozen or so names inscribed on it. Beside each name there was an electric buzzer and the floor number. Monsieur Claude Géraudel lived on the third floor. I pressed the buzzer. In a little while a bell rang, and only then did the second door budge, opening onto a staircase. I struggled up it with a heavy suitcase in either hand.

On the third floor the apartment door was already open to reveal a man with grey hair and grey eyes, and an astonished expression.

The only times I had seen a chief administrator or an overseas governor were in the municipal stadium at Lomé during the Joan of Arc celebrations or the July 14 festivities, which for us were grand events with a parade lasting hours. We would sing to the glory of the 'Motherland' as our marching feet churned up the dust. The 'Marseillaise' would ring out; the administrator, standing rigidly at attention on the rostrum surrounded by an impressive array of officials and notables, would take the salute. The blazing sun would burnish the buttons of his white jacket and shimmer on his medals and the gold braid on his cap. At the time I would never have believed that one day I would climb the steps that separated me from such an exalted personage.

But the man now standing in the open door was wearing a plain, light, charcoal-grey suit. Not quite sure if I had found the right person, I asked:

'Is this where Monsieur Claude Géraudel lives?'

'I am he,' he replied in surprise, with one eye on my luggage.

'I have come at the suggestion of Monsieur Guy Echevrery, the technical counsellor at Dakar,' I said, handing him Guy's letter.

'Come in,' he said, after a glance at the handwriting.

Taking one of my suitcases, he showed me into the drawing room, waved me to a chair, and sat down opposite me in a chair with a high back. Daylight was streaming in through three tall windows: the light was softened by the fine lace curtains and glowed gently on the parquet floor, with its rich oriental carpet woven with wonderful foliage patterns. A caged parrot was swinging on his perch, squawking and flapping his wings: his harsh voice seemed to lend life to a forest landscape whose large gilded frame shone dimly at the other end of the room.

At first we talked about Guy, our common friend, then about my travels.

'Well now,' said my host, putting the letter back in the envelope, 'so you want to link Greenland with Africa. That's an unusual ambition for an African. Coming from a country which scarcely three years ago was still a French colony, you are well placed to compare what we were able to accomplish during half a century in Africa and what has been done since then for the Eskimos. I am convinced that we managed to achieve some good things, if only in the field of literacy. And you are living proof of that: not only do you express yourself in correct French, but you also show great open-mindedness towards the outside world – the result of a remarkable education.'

I felt flattered. My host didn't yet know that I had taught myself.

'But tell me how you manage to get along,' he went on. 'Have you a grant from your government or from some organization? Ah, what a pity you have not yet been to a university, for then it would have been easy to get you a scholarship.'

These two words, grant and scholarship, kept recurring all the time in my conversations with others. Each time I was astonished to learn that in our day it's thought odd to set one-self a certain goal without asking for monetary help. As for university studies, it seemed to me that without them I had remained more African: an African graduate would have spurned the risks I took and found a niche in some government ministry.

'Your example deserves to be made known to all young people in Africa. You also deserve help. I'll be leaving soon for a few months in Morocco, but until then you're welcome to stay here with me.'

He showed me round the apartment, then took me to the room that was to be mine. My host had been an administrator in Guinea, and after that in Dahomey. Since the independence of those two countries and his early retirement, he divided his time between his sumptuous apartment in Paris and pleasure trips to Africa, where he still had many friends.

That evening he took me on the Métro and showed me the lights of the Place de la Concorde: I was enchanted. Then we strolled along the Champs-Élysées as far as the Étoile, a considerable distance on foot. But my host was agile despite his age. Tall and slim, his face seamed by the burning sun of Africa, he walked along with great regular strides as he explained the history attached to each monument. That was how I learned that the Place de la Concorde did not always have the name it bears today. Halfway between that great square and the Étoile, I responded like a child to the fantastic neon signs which lit up the shops and cinema fronts. To me it was all like some fairy tale where a magic wand gave an enchanted look to everything I saw. We dined in a restaurant on the Champs-Élysées before going home. That evening, sliding between fine linen sheets, I forgot the long sleepless nights spent in the ports and 'bush taxis' of Africa.

About three months after my arrival, Monsieur Géraudel had to get ready for his journey. I still had not managed to find lodgings for the few months before I myself left Paris. Then, a week before leaving, he had a guest for dinner named Monsieur Jean Callault, an old friend. Slightly stoop-shouldered and quite stout, he was just the opposite of my host, hobbled by the chronic arthritis from which he had been suffering for some fifteen years. Because of a bone fracture resulting from a fall, his left shoulder was a little higher than the other. This gentleman made up for his slight physical defects with a tender, responsive heart. His soul shone through his clear blue eyes, and his thick eyebrows never knitted in a frown except to express sorrow and compassion for the sufferings of others. He was a brilliant talker and could recite by heart, well past his fiftieth birthday, whole pages of Racine. I remember the recording we made on a cassette tape in which he alone played the roles of both Phèdre and Hippolyte, then of Bérénice and Titus, without once having to consult the text. 'I should have been an actor,' he used to say; as a young man he had indeed thought of becoming one, but his parents had pushed him into accountancy. Not that it now prevented him from keeping up with all the developments in the theatre, fashion, in fact everything that happened in Paris in the fields of literature and art, or from being acquainted with a good many artists. A universal spirit, he showed a keen interest in me and in my plans. When he heard about my accommodation problems, this old fellow who clung fiercely to his bachelor freedom offered me hospitality.

I moved four days later. My new host lived in a luxurious apartment in the seventeenth *arrondissement*. As soon as you entered the salon, with its polished, gleaming parquet, its sofas upholstered in velvet, low table covered with meticulously arranged periodicals, and glass-fronted bookcase, you could not help being struck by the strict order that prevailed around this almost eccentrically tidy man. In my bedroom (whose two windows overlooked the quiet rue Philibert-Delorme) there were three large cupboards accommodating the overflow of books from the library. All the books were covered with dust jackets, then placed in envelopes or folders and laid on top of one another. The books were piled right up to the top of the cupboards. Unfortunately there were no labels on the packages, and sometimes Monsieur Jean Callault would spend a whole morning looking for one particular book. Perched on a chair, he would keep opening and closing the folders, repeating: 'And yet I'm sure I put it . . .' He had been accumulating 'reading matter' for over twenty years in anticipation of his retirement, which was due any time now, or so he said. However, ten years later (during which time I kept up a correspondence with him) he was still looking forward to that mythical event.

I stayed for eight months with Monsieur Jean Callault and owe him a deep debt of gratitude, not only for the comfort and security he gave me, but also and above all for his great kindness. He became a real father to me, as we shall see.

When one has been made as welcome as I was, it is hard to agree with the common sweeping judgment that the French are not very hospitable. Certainly, like anyone else, I ran into occasional trouble in a city as large as Paris; but this was amply compensated for by the two men whose goodness of heart and simplicity of manner made me more optimistic than ever. It was in this state of mind that I continued my journey across Europe.

I caught the train for Bonn in February 1964, after working in a warehouse belonging to the Au Printemps department stores in Saint-Denis. I had been told before I left that, as a Togolese, I didn't need a visa for Germany.

At the Belgian frontier, however, I was asked for a visa for Belgium.

'Germany is further on,' said the police officer. 'You won't get there till later. In the meantime you have to cross Belgium.'

'In that case, I'll go straight through.'

'You need a transit visa. I must ask you to leave the train.'

I was at the Belgian police post at Erquelinnes. A stamp was banged down on a page of my passport and the officer wrote on it: 'Sent back, no visa.' The train left me at the frontier on French territory, in front of the station at Jeumont. As the focus of attention for the other passengers staring impassively out of the windows, I tried to put a brave face on things, too baffled even to feel upset. As the train departed I continued to stand there smiling, as if in a dream.

It was nearly midday. What was I to do? Wait for another train and return to Paris? From Jeumont I telephoned Monsieur Jean Callault

'I warned you that you'd have difficulties on the way. If you can't go on, come back here.'

I was feeling inclined to follow his advice, when the station master informed me that there was a Belgian consul at Hautmont, near Maubeuge. A single taxi served the area.

'If you hurry,' the station master added, 'you could still catch the afternoon train for Bonn.'

I followed his advice and was allowed to cross Belgium.

I arrived in Bonn at about eleven o'clock at night. As I had no letter of introduction this time, I was planning to find a hotel when something happened at the station.

Among the passengers who had got off at Bonn were two ladies – or rather, a lady of about forty and a girl – who hadn't found a porter and were walking along the platform struggling with their four suitcases. They were literally dragging two of these along: they would go two or three steps with them, set them down, then go back and drag the others, and so on. They even seemed to be getting some fun out of it. But they also had some small parcels in their arms, which didn't make their task any easier.

The first thing I did was to leave my luggage in a corner, then I went up to these two charming ladies and offered to help. Although surprised, they accepted, and I took their suitcases out to the street, where they hoped to catch a taxi, before going back to get my own luggage.

When I came out again onto the pavement with my own two big suitcases, there were still a lot of people waiting for taxis. I had already dismissed the previous incident from my mind, and probably wouldn't have seen the two ladies in the crowd if one of them hadn't caught sight of me and said:

'Young man, that was very kind of you.'

I replied quite simply:

'Yes, I'm kind.'

The words just slipped out, but they had a most unexpected effect. The two ladies burst out laughing and then, thinking that they were dealing with someone who couldn't speak their language well, one of them asked:

'Perhaps you haven't been long in Germany?'

'I've just arrived, and shall be staying a few months before moving on. I'm going to the North Pole.'

They gaped at one another. Then two pairs of bright blue eyes were levelled at me. No description would do justice to their astonishment. Those two little words 'North Pole', coming from the mouth of an African, had almost knocked them over.

'Well,' I said, trying to reassure those two pleasant ladies, 'I'm going to Greenland to live with the Eskimos.'

'But, *mein Gott*, you'll freeze to death there!' said the elder of the two, instinctively putting her hand to the collar of her fur coat.

It was now my turn to show surprise. Her gesture made me understand for the first time the meaning of the word 'freeze.'

'Is your country's government sending you there?'

'Not at all! The government of my country has other things to worry about.'

'What country do you come from?'

'From Togo.'

'Togo? *Ach so! Eine alte deutsche Kolonie vor dem Ersten Weltkrieg!*' (Oh, yes! A German colony before the First World War!)

'*Ja.*'

'But that was long ago. How come German is still taught there?'

'There's a Goethe Institute in Lomé.'

But all the time we were talking about my country, I was growing more worried about that word 'freeze'.

'What does it matter!' I said aloud. 'Freeze or not, I'm going there all the same. I left my land a long time ago to go and live with the Eskimos.'

My companions finally convinced themselves that they weren't dealing with a practical joker, so our conversation became more and more interesting as we made our way to the head of the taxi rank. The elder of the two ladies asked me to tell them about my travels and the various stages of my journey, and I did so with enthusiasm. The younger, whom her companion called Carola, was less talkative: she simply kept looking at me and smiling. There were only three people in front of us when the older lady asked me:

'Have you somewhere to stay in Bonn?'

'I'm going to look for a hotel.'

It was almost midnight. The lady turned to the girl and asked her:

'But . . . we can easily put him up, can't we, Carola?'

'Of course,' she replied.

69

So they offered me a place to stay.

The elder lady was Frau Anna Sprick, a widow, and young Carola was her cousin.

I lived for a year in their pretty apartment in Münsterstrasse, not far from the station. That year I worked in a fizzy drinks factory in Roisdorf, a suburb of Bonn, preparing to leave for Denmark, my last stop in Europe.

Last-minute difficulties cropped up in Copenhagen. The Danish commissioner who was supposed to issue my visa for Greenland was confronted with a delicate situation. He felt it was a great personal responsibility to allow a young African to set off under his own steam for that land of ice: he was uneasy in his mind about it.

'What's the temperature in your country?'

'An average of thirty-five degrees.'

'In Greenland it's forty degrees below. For you, that's a difference of seventy-five degrees. You'll have trouble adapting. Unless you're going to experiment with hibernation? But you'll still have to wake up when the winter's over!'

He stressed the dangers and tried to make me change my mind. Though my visa for Denmark was also officially valid for Greenland, a dependency of the Danish kingdom, the commissioner informed me that I also needed special authorization to sail there.

'We'll let you know by letter,' he concluded.

At the ministry concerned with Greenland, I met no such systematic discouragement, but rather a disconcerting eagerness to help. I was welcomed everywhere and given every kind of information. I think they were secretly convinced that I'd never leave the charming city of Copenhagen to bury myself in the icy wastes of Greenland. Then the director of cultural affairs at the French embassy, Monsieur Révil, intervened on my behalf with the minister for Greenland.

My case presented no real problems, except to the commissioner. Since arriving in the city I had been living at the Central Hotel, near the station, expecting to stay for only a few days, but three months passed and I still had no authorization to leave for Greenland. What were three months, though, after so many years of perseverance? I hung on and took a job as a dishwasher at the Frascati, a big restaurant near the City Hall which specialized in French cuisine. It was a judicious choice, because my meals were free. So I managed to meet my hotel bill every week and put aside a few hundred kroner a month, while waiting for the commissioner to overcome his scruples. My job at the Frascati consisted of nothing but washing large beer glasses. Indeed, the customers who sat on the terrace from ten in the morning till late at night polished off a prodigious amount of that good Danish drink every day. My free time and my days off were spent in the National Library or the National Museum, which had an important collection of works on Eskimo art. I could already understand Danish: only the rather daunting pronunciation prevented me from articulating correctly.

I was keeping up a regular correspondence with Monsieur Jean Callault, the one friend I still had from the previous stages of my journey. His letters, full of cautionary advice, were like letters from a father to his son. I entered wholeheartedly into this increasingly affectionate relationship, writing to him as to my own father, and so perhaps making up for the guilt I felt at leaving my own family without news of me – a family so long neglected, and which would now have difficulty in understanding me. Soon my letters to Monsieur Jean Callault began, 'My very dear papa'.

In April, my adoptive father (as I shall now call him) invited me to spend a few days in France at his expense before leaving Europe. I spent Easter week with him in Paris, then returned to Copenhagen, where I was still having difficulties with the commissioner.

'How will you live there?' he fretted. 'Have you enough money?'

So my adoptive father came to Copenhagen and stood guarantee for me, and the long-awaited authorization was finally entered and legally endorsed in my passport.

'Don't get crushed by an iceberg,' said the good commissioner by way of farewell.

Eight years had passed since my departure from Togo.

PART II

The Call of the Cold

A Spirit from the Mountains

My final preparations had been simple. A stroll around Nyhavn, a picturesque district near the harbour, enabled me to pick up an old pair of American army boots at a bargain price, an overcoat with a quilted lining, two woollen pullovers, and two pairs of mittens. This was the extent of the equipment I assembled to answer the call of the north. I suppose I was travelling light. My adoptive father had presented me with an ancient folding camera that he had owned for a quarter of a century. Finally, I bought some paper for a diary. All this was squashed into a rucksack.

I had decided to travel by ship: it would be rash for someone like me suddenly to come up against intense cold after only a few hours' flight, whereas a sea voyage of several days would allow me to adapt gradually to the climate. Quite a sensible idea, coming from one so often accused of lacking common sense.

The *Martin S*, a cargo and passenger boat, had been in port since May, loaded and ready to make her first voyage of the year. She was bound for Julianehåb, one of the first towns in southwest Greenland. I bought my ticket for that destination.

In fact, my plans were almost as ramshackle as my frail equipment, for though I knew I wanted to live with the Eskimos, I still had only the vaguest idea as to just where I should stay in that vast land. I decided that if I landed in the far south I could then make my way northward up the west coast and so live in several townships.

That Greenland is the biggest island in the world, with an area of 2,175,600 square kilometres, is merely abstract knowledge giving no idea of its real size. It takes a few comparisons to give a more solid impression. The distance from Cape Farewell, the southernmost point in Greenland, to Cape Morris Jessup in the north is the same as between London and the mid-Sahara. Its breadth is equal to the distance from Paris to Copenhagen. It is an immense desert with only thirty-five thousand inhabitants. Only the coast, consisting of rocks and high mountains indented with deep fjords, is habitable. The interior (five sixths of the total area) is entirely covered by the *inlandsis*, or continental ice, which can be as much as 3,200 metres high and 3,500 metres deep. If it were all to melt, the world's oceans would rise ten metres and drown whole coastal cities. Because of the huge mass of this glacial cap, we still don't know if Greenland is really an island, or an archipelago covered with eternal ice.

The ship left port on the afternoon of June 19 in foggy, dreary weather. At the dock ten people, all warmly clad, waved goodbye. Soon Copenhagen had disappeared and I began to feel closer to the Far North. But it was just the start of my adventure.

There were only nine passengers. An easy fellowship developed, and an excellent atmosphere prevailed on board right from the start of our journey. Among the eight other passengers were two Danish women (a mother and daughter, who were going to visit a relative), a Greenland woman and her child, and a pastor. There was also a young Dane called Chris, a skilled construction worker, who was taking his craft to the Eskimos 'to help them live better in more modern houses'. Adam, a Greenlander of thirty-two, had been working as a cook in Sweden and was going to spend his holidays in his native land after a twelve-year absence. His wife, a Swede, and his daughter of eight had stayed behind. But the passenger with the most surprising mission was a young Greenland woman called Tuperna, who had been taking a one-year course at an institute for beauticians in Denmark. Now she was returning to Narsaq, her native village, to open the first hairdressing salon!

It was not long before I had the first surprise of the trip. On our first day at sea, the ladies sunbathed until nine o'clock in the evening; on the second day, until ten o'clock; and on the third day, until eleven! After we had said good night on deck, I would go down to my cabin and read by the light of the sun. The brief, pale 'night' that followed (night only in name) would soon fade like mist. Towards three in the morning, it was daylight again; bright sunshine filled the cabin with its warm red rays. The short night grew shorter day by day as we neared Greenland. 'If it goes on like this,' I told myself, 'there'll soon be no night at all.' Every two days, when we got up, the ship's clocks were put back one hour in accordance with the changing time zones. I was astonished to find that it's only nine o'clock in the evening in Greenland when it's midnight in Paris and eleven o'clock in the evening in my native land. That set me thinking: with the total absence of night for six months in Greenland, how much sleep did the inhabitants get? For the time being, this question seemed insoluble. As for myself, I had already lost all track of time and never knew quite when to go to bed.

Right from the start, the fine weather promised a real pleasure cruise. Every morning a charming Danish hostess woke up the passengers for breakfast. Like guests in a hotel, we strolled along to the dining room, one by one. The white tablecloths, with their rows of napkins either knotted or folded like bishops' miters, gave the room a festive air. After lunch, we would retire to the smoking room and then go to our cabins for a siesta. The whole interior of the cargo boat was attractive, and each of the cabins, matching the rest with their walls of varnished pine, had its own toilet and shower. In addition to the bunk, there was a divan, two armchairs, a low table anchored in the centre, and olive-green carpeting on the floor. The *Martin S* was new; this was only her second voyage.

On the fourth day, we were still able to lie out in the sun until eleven at night, but that pleasure disappeared on the following day. The sea turned rough for the first time, alternately worsening, then calming again. Finally a gale blew up and kept on blowing. The sea swelled, and the waves raged and thundered over the decks. Doors kept slamming shut. Everyone kept to his cabin. I rushed into mine to vomit, then lay down on the bunk. The other passengers were throwing up into the scuppers. I couldn't find a comfortable position on my bunk: its angle was determined by the ship, which kept pitching and tossing. That lasted all day. In the toilet next to my cabin, the water in the toilet bowl rose and fell with the rolling movement of the boat; suddenly dropping out of sight, it would shoot back without warning and spurt up to the ceiling. When you went to leave the toilet, the alarming angle of the floor prevented you from stepping forward; then the boat would slowly roll in a sickening lurch that hauled on the door and snatched the handle from your grasp. The door would swing wide open to let you through before shutting with an ear-splitting bang, and you would find yourself propelled towards your bunk, which broke your fall.

We were a handful of human beings far from land, imprisoned in a ship that plunged and reared and was dwarfed by mountainous waves. The thought of sinking nagged at us all, though no one mentioned it. None of my previous voyages had reduced me to such dark despair.

The 'night' of Wednesday, June 23, was extraordinary: a night of wild commotion. Objects kept tumbling over with tremendous crashes, and luggage ploughed from one end of the cabin to the other, while the whole vessel creaked and groaned. Next morning the sea suddenly became as smooth as a millpond. The air turned so cold that it started to cramp my breathing: we were approaching the barrier ice. Towards one o'clock in the afternoon, we spotted the first ice floes.

These were ice blocks of varying shapes and sizes, drifting here and there as the waves took them. The smallest looked like swimming swans, and some were like crouching camels rocking gently from side to side. Some were white, others green or blue. A brilliant sun, cold as steel, glittered on them and transformed the sea into a fairy-tale world: a vast ice-blue expanse strewn with great chunks of crystal. A dazzling glitter seethed and multiplied.

Half an hour later, these blocks were the size of anthills, and the much larger submerged parts formed enormous azure masses in the glaucous depths as they passed us. A little later on, they had reached the height of hills: these were the fabled icebergs, transported by the polar current that flows along the coast. Soon we could see hundreds of them, and their size never ceased to astonish me.

Meanwhile the smaller masses were becoming more and more densely packed. By evening, we could see only one vast sheet of ice spread all across the sea, with white mountains rising here and there. Creeping through this pack-ice, the ship left a narrow channel in her wake that quickly filled again with slabs of ice. Her sides became colder and colder. Then came dense fog, and we stopped in the midst of the ice during the brief night, awaiting daybreak and better visibility.

Next morning, Saturday, June 26, the sun rose at two o'clock, shining so strongly that it dispersed the fog and we were able to move again. The ship resumed her slow progress, interrupted by bumps, stops and starts. Your hands went numb as soon as you took them out of your pockets; a raw cold bit at your ears,

79

your nose; your face felt frozen to the touch. My breathing became more and more laboured, and painful stabbing sensations in my nostrils made it agony to inhale deeply in that frosted atmosphere. I put on my thickest pullover, then my woollen mittens, and, as the soles of my American army surplus boots weren't thick enough to keep out the icy chill of the decks, I wore two thick woollen socks on each foot.

That Saturday, after seven hours' endurance and hard navigation, we sighted land: Cape Farewell, the southern tip of Greenland! Mountains rose in the distance behind a slight mist, and the sight gave us all fresh hope.

But the captain, who usually wore a broad smile, radiating confidence, had a grim look and stuck to the bridge. This was because many vessels had been shipwrecked near Cape Farewell, after riding out storms: the icebergs still lay in wait. I still remember the party the captain organized on board on the evening of our arrival off Julianehåb, to celebrate the successful end of our journey. (The *Martin S* was shipwrecked the next year, while I was still in Greenland.)

We arrived on Sunday, June 27, towards noon. The sun seemed much warmer when the ship finally emerged from the immense sea of ice and sailed up the fjord to Julianehåb. In front of us, the bare, snow-capped peaks of tall grey mountains were silhouetted against blue sky. Some were haloed in mist. In the fjord, a few small icebergs were floating in the tranquil water.

Julianehåb is still called Qaqortoq, 'the White One'. This Eskimo name, the image of the wilderness of white, was given it because the masses of ice and the icebergs that drift down both coasts pile up in this area. Embedded in pack-ice a hundred kilometres wide and from three to ten metres thick, they prevent ships from reaching the east coast for months – sometimes for ten months at a time.

When we had sailed up the fjord, the first sign of Qaqortoq was a collection of about thirty little houses made of wood and painted yellow, green, blue or red. They were scattered at the foot of a huge mountain, in the midst of green lichens looking as soft as a lawn and sprinkled with yellow flowers. This tundra, the sole vegetation clinging to the rocky soil of that Arctic land, casts an irresistible spell over the traveller who has just spent so many days at sea.* The great mountain hid other houses which became visible after the ship had entered the harbour – a sort of wooden landing-stage with a warehouse of brick and stone. Qaqortoq (from here on I shall call it by its Eskimo name) consists of about three hundred and fifty houses, all made of wood and horribly alike, perched on the flanks of huge rocks or scattered along the valleys. There are eighteen hundred inhabitants, making it after Godthåb, or Nuuk, the capital, which itself has only five thousand inhabitants, one of the most populous 'towns' on the island.

We pulled alongside the landing-stage, and through the porthole of my cabin I could see the entire population gathered on the square beside the warehouse. The men were of shortish build, though they were mainly half-castes, and clad in thick cloth trousers and pullovers or anoraks. The plump women wore European overcoats that came down to their ankles and headscarves and *kamiks*, or sealskin boots. They were holding an impressive number of chubby little children by the hand, their stout young bodies almost bursting out of their clothes. All stood silently watching the ship which, after an eight-day voyage – two of them spent battling against the barrier ice – was at last arriving in their village, bringing so many long-awaited

* On disembarking around AD 983 near Narsaq, not far from Julianehåb, the Viking chief Eric the Red called this place Grøn-Land, Green Land, which became Greenland in English and Groenland in French. A land without trees, but green with lichen.

81

goods, especially – yes, especially – coffee, tobacco and alcohol! But I mustn't get ahead of my story.

Never were inhabitants of a mountainous land more peaceful looking. They kept smiling, exchanged admiring comments on the *Umiarsuaq*, the big ship, and laughed openly at the awkward walk of the passengers, who were staggering about the deck like convalescents.

I wondered what their first reaction would be on seeing me, a black man, leave the ship. They had never seen a man of my race, except perhaps in newspaper photographs. Like an actor carefully preparing for his first entrance, I took my time dressing in my cabin, putting on a thinner pullover under my overcoat. Unhurriedly I drew on my woolly mittens, pushed back the hood of my overcoat, and then, with my hands in my pockets, I made my entrance.

As soon as they saw me, all talking stopped. So intense was the silence, you could have heard a gnat in flight. Then they started to smile again, the women with slightly lowered eyes. When I was standing before them on the wharf, they all raised their heads to look me full in the face. Some children clung to their mothers' coats, and others began to scream with fright or to weep. Others spoke the names of *Toornaarsuk* and *Qivittoq*, spirits who live in the mountains . . . That's what I was for those children, and not an Inuk* like themselves. Like children the world over, they spontaneously spoke their minds about me. Unfortunately, I can't say the same for the adults. Proud and secretive, they masked their feelings behind an unchanging

---

* *Inuk*, plural *Inuit*, the name by which the Eskimos refer to themselves and which is applied to them only. It signifies 'true man'. To them, other races are merely the result of the coupling of a woman and a dog. Like all other Eskimo peoples, the Greenlanders don't know the word 'Eskimo'. This appellation, which comes from the Algonquians, neighbours and enemies of the Eskimos, is a term of contempt. It means 'eaters of raw flesh'.

smile, mild but enigmatic. Not one of them corrected the children, yet the mothers' calm gave some of the children confidence, and, as they saw me approaching, they too tried to smile – a hesitant, not very reassuring smile.

The crowd opened to let me pass. It was then that I distinctly heard a woman speak the word *kusanaq*, a flattering term that I didn't understand at the time, but which means 'handsome'. Handsome in what sense? For the children I was a fearsome supernatural being who came to exterminate the village. Besides my being black, it must have been my height – five feet eleven (1.80 m) – that contributed to the fear I inspired in the children, whose parents were little more than five feet three (1.60 m). Perhaps too for that woman, a small woman living with small men, I was handsome precisely because of my height. So my height impressed them, but in different ways according to their age. It spread terror in the children, astonished the men, and was attractive to one woman who at that moment was probably summing up the opinion of all the other women. Two days later the radio station in Godthåb, the capital, announced the arrival of an African in the country in these terms: 'He is a very tall man with hair like black wool, eyes that are not slanting but arched, and shaded with curling eyelashes.'

The new employer of Chris, the young Danish builder, took us to his house for a beer. All the children left their parents' sides and followed us. More of them came out of every house; soon I had such a procession of them behind me that Qaqortoq's scanty police force was obliged to follow us in a patrol car moving at walking pace to keep the children from trampling the 'foreigner' underfoot. The scene made me think of the Lilliputians surrounding Gulliver. I had started on a voyage of discovery, only to find that it was I who was being discovered.

Adam was welcomed by his parents.

'I'm going to find you somewhere to stay,' he said as he left us.

He came back to see me at the home of Chris's boss and announced that everybody wanted to put me up! But when his own sister, aged twenty-eight and married, heard that he was looking for lodgings for me, she absolutely insisted on having me stay in her house.

We walked past the little wooden church. Soon the house came into view. Like all the buildings there, it was isolated, standing about fifty metres apart from the other houses. Because of the uneven ground (formed of bare rocks worn down by glacial erosion) and a stream running nearby, the house stood on a solid foundation of brick and stone. Three wooden steps led up to the entrance.

We found ourselves in an unheated entrance hall about two metres long, with clothes hung up on nails. Black rubber boots were strewn across the floor, and I could see plastic buckets lined up along the wall, next to a barrel. Without knocking, Adam opened the second door, which gave access to the living room. A little woman, all smiles, came to welcome us.

' *Velkommen,*' she said in Danish.

'*Tak.*' (Thank you.)

We settled down in comfortable armchairs.

'Cigarette?'

'No, thanks, I don't smoke.'

She lit one herself.

Green plants and pots of geraniums decorated the window ledges. On a low table was a combination radio and gramophone. A big sealskin was spread on the floor, the animal's claws gleaming between the legs of a sofa that filled a quarter of the room. The wooden walls were plastered with family photographs, pictures of Jesus, and large coloured portraits of each member of the Danish royal family.

My hostess dragged heavily on her cigarette.

'Does your friend drink coffee?' she asked Adam.

'*Aap*,' he replied.

The young woman at once brought in some cups and a coffee pot from the kitchen. I was expecting Nescafé, but this was real fresh-ground coffee. Its mellow and penetrating aroma took me by surprise.

'How could she have made coffee in less than a minute?' I aske l Adam.

'In all the houses here, the coffee pot stands ready on the stove in case of visitors.'

Within a few minutes, we had each drunk five cups of coffee served with biscuits. After the third cup, she poured a few drops of *akvavit* into the coffee. The talk became lively, and Adam drew on all his Danish and English as interpreter.

'Mikili' (Michel), said my hostess, turning towards me, 'my name is Paulina. I've told Adam that you will stay here with me. Is that all right?'

'Fine.'

'You can bring your luggage this evening. Come and see.'

She took me upstairs to the second floor, where there were two rooms facing each other.

'You'll take that one,' she said, pointing to the room on the right.

It had a metal bed covered with white sheets, clean but not ironed, and an eiderdown. There was a chair beside the bed. That room was Paulina's and her husband's, but before our arrival in the house Paulina had moved out their belongings and put them in the children's room on the left, so as to give me the better of the two rooms.

'Hanssi and I will sleep in the children's room opposite,' she said.

'What about the children?'

'They'll sleep on the floor.'

I protested, especially since Hans, Paulina's husband, who

CAPE MORRIS JESUP

ARCTIC OCEAN

GREENLAND

Siorapaluk *(Etah)*

Qaanaaq *(Thule)*

NORTHEAST
GREENLAND
NATIONAL
PARK

Pituffik *(Dundas)*

Store Koldeway

Savissivik *(Savigsivik)*

Daneborg

BAFFIN BAY

Kullorsuaq *(Kuvdlorssuak)*
Nuussuaq ● Nutaarmiut *(Kraulshavn)*
Tasiusaq *(Tasiussak)* ● Aappilattoq *(Augpilatoq)*
Upernavik ●
● Kangersuatsiaq *(Prøven)*
Upernavik Kujalleq *(Søndre Upernavik)*
Igdlorssuit *(Illorsuit)* ● Nuugaatsiaq *(Nûgâtsiaq)*
Uummannaq *(Umanaq)* ● Ikerassak
Qullissat *(Qutdligssat)* ● Sarqqaq *(Saqaq)*
Qeqertarsuaq ● Appat *(Ritenbenk)*
(Godhavn/Disko Island) ● Oqaatsut *(Rodebay)*
● Ilulissat *(Jakobshavn)*
Aasiaat *(Egedesminde)* ● Ilimanaq *(Claushavn)*
● Qasigiannguit *(Christianshåb/Christianshaab)*
● Kangaatsiaq *(Kangâtsiaq)*
Sisimiut *(Holsteinsborg)* ●
● Kangerlussuaq *(Søndre Strømfjord)*

Ittoqqortoormiit
(Scoresbysund)

Maniitsoq
(Sukkertoppen)

● Kulusuk *(Kap Danmark)*

Tasiilaq *(Ammassalik)*

Nuuk
(Godthåb/Godthaab)

Kangerluarsoruseq
(Færingehavn)
● Qeqertarsuatsiaat
(Fiskenæsset)

ATLANTIC OCEAN

Paamiut
(Frederikshåb/Frederikshaab)
Ivittut *(Ivigtut)*
Arsuk ● Narsaq *(Narssaq)*
Qaqortoq *(Julianehåb/Julianehaab)* ● Igaliku *(Igaliko)*
Nanortalik *(Nanortalik)*

CAPE FAREWELL

Key: **Greenlandic names**
(Danish names)

0   200   400   600 km

was out at work in the naval dockyard, had not been consulted. But Paulina would hear no arguments.

'Anyhow,' she said, 'Hanssi is nearly always drunk, so he might as well sleep on the floor. Do you know what *immiaq* is?' she asked eagerly when we were back in the living room.

'No.'

She disappeared into the kitchen, brought three glasses, and filled them with a yellowish liquid that she scooped in a bowl from the barrel I had seen in the entrance hall.

'This is *immiaq* – Greenland beer.'

I took a little sip from the glass that she had filled to the brim. The drink had a sour taste that reminded me vaguely of dry cider.

'*Mamarpa?*' (Is it good?) she asked rather anxiously.

'*Aap, mamarpok,*' I said.

We went on alternating cups of coffee and glasses of *immiaq*.

'Adami,' Paulina said briskly, 'go get some food.'

Adam brought a dish full of big slices of some sort of meat from the kitchen and deposited it on the round table.

'This is *mattak*,' Paulina told me.

'*Mattak?*'

'Yes – raw whale skin.'

The slices of meat were as thick as papaya pulp or the flesh of a melon. We each took a slice and clamped it between our teeth; then, holding the other end in one hand, with a knife we cut off a bite, Eskimo fashion – that is, cutting upwards close to the lips, so that you run a great risk of slicing off your nose in the process.

Mattak consists of two adjoining layers; the outer part (the natural skin of that type of whale) is of a matte white, has the firmness of cartilage, and is rather tender, soft, even succulent, while the pinkish inner layer is on the contrary very hard to chew.

This new diet caused me considerable alarm. I wondered if I

was to be fed on nothing but whale skin during my stay in Greenland. I could still change my mind if I wanted to. The ship, which was to call at two other ports before returning to Denmark, would be staying for several days at Qaqortoq. It crossed my mind to go straight back to Europe, but I hesitated. Could I suddenly give up what had taken me so long to achieve, just because of a bit of raw whale skin?

So I ate my portion of whale skin, and my hospitable hostess asked me if I liked *mattak*. Fear of disappointing her made me reply, 'Oh, yes!' Whereupon she gave me, all to myself, a whole plateful of the stuff, to which she added an enormous quantity of yellowish, blood-tinged seal blubber. Very slowly and with great difficulty, but smiling appreciatively, I managed to finish the larger part of this food, which was not made palatable by any spices, not even salt. Inwardly, fearing a frightful stomach ache, I berated myself for my excessive politeness, but without it I couldn't have won the precious friendship of my hostess so quickly.

Paulina went back into the kitchen, where she had so many surprises in store for me, and brought out some *ammassat*, little dried fish resembling skinny herrings. These salmonidae, some-times known as caplins, are also eaten with seal blubber. I found them better without that delicacy, but Adam and his sister disagreed.

'Here,' they told me, 'you must eat plenty of fat – it helps keep out the cold.'

I left Paulina's house in the late afternoon, accompanied by a belching Adam.

'Bring your luggage this evening,' his sister told me again.

All in all, I was pleased with that first visit. The welcome couldn't have been warmer. I went back on board the ship duly escorted by children and police. All this time, in fact, the chil-dren had been thronging around the house.

The first thing I did was to go see the captain again and ask if

Arriving at Qaqortoq (Julianehåb)

I could continue to eat on board during the few days the *Martin S* would stay in port.

'At least once a day,' I pleaded. 'To vary my diet.'

He granted me this favour, and even added that I could use my cabin, too, for two or three days.

Like a swarm of bees around a hive, young girls wearing brightly coloured anoraks and tight-fitting jeans invaded the ship, going into the Danish seamen's cabins. I could hear one girl saying outside on deck: 'But where's the cabin of Mikilissuaq (Michel the Giant)?' Then one of them came in.

A few moments later she was fast asleep on my bunk. The door opened again a bit later, and a smiling little man with a rebellious lock of hair dangling over his forehead came into the cabin, accompanied by Adam. It was Hans, Paulina's husband.

'*Illumut!*' (Let's go home!) he said.

The girl woke up and protested:

'*Naamik!* He's going to stay with me. Aren't you, Mikili? I live alone with my father.'

'But my wife has already made his bed!' Hans retorted furiously. 'Mikili, *illumut!* My children will come and fetch your luggage.'

I tried to explain to the charming, disappointed girl that Hans's wife really had offered me hospitality already. I had accepted and I had to keep my word.

So I let myself be hauled off by Adam and his brother-in-law. They didn't take me home right away: we went for a while to chat in the 'café-bar'. There, young mothers were drinking beer, with blond children sitting beside them or on their knees – the consequences of too many visits to the boats . . .

I returned to the ship only to attend the captain's party. There I met some of the local Danes. Apart from girls, no Greenlanders were invited. The drink flowed like water. Chris spent the night in his cabin with a young native woman who was three months pregnant. As for me, I left the ship a little before the party ended. Hans was waiting for me, sitting on a stone near the warehouse, and guided me to his house.

Such was the welcome I received on the first day of my long stay on the western coast of Greenland.

2

Queer Customs

Hans's house, painted yellow, stands on a rocky eminence at the far end of Qaqortoq, between the main street leading down to the naval dockyard and the stream whose source lies in the high mountains covered in June by a cloak of snow. This stream crosses the village and runs into the fjord.

My room had an uncurtained window opening onto the main street that ran past it like a long ribbon. The very uneven ground, whose bare rocks were rounded like tortoise shells, their outlines marked by tufts of grass, sloped down to the bank of the stream. So the view from my room overlooked the scattered houses, and I could see straight into the fjord where icebergs floated. An intense light slanting onto the houses, casting long shadows and reddening the tundra, showed that it was early morning. A morning after a night without darkness. Apart from a few seagulls hovering over the fjord, nothing moved in this strange morning light. The silence was overwhelming.

Suddenly, near the one and only bridge, a man in a thick pullover and with *kamiks* on his feet left his house. He closed the door gently, walked down the wooden steps, and paused, hands in pockets, facing the fjord. For at least five minutes he stood there motionless. The legs of his thick trousers were bunched up over his *kamiks*, and the broad folds on the arms and the chest of his pullover gave him a tough, stocky look. For a good half hour this man kept walking up and down in front of his house, apparently aimlessly. Surely he wasn't just taking a morning stroll – at least not in the usual meaning of the phrase. His pacing to and fro was strictly between his house and the street, which he neither crossed nor walked along. Then I saw him walk four or five times round his house at the same pace, still with his hands in his pockets and with his head jutting forward. Perhaps the man was just bored, I thought. Now he was standing at the edge of the road, his face turned towards Hans's house. His face was half hidden by a straggly beard, and long black hair hid his ears and fell straight to his shoulders – he looked like Robinson Crusoe! His expression changed, and he gave a little smile. Could he have seen me at this distance? I waved to him. He waved back, then began walking up and down again in the bright sunlight.

A little later, a second man left his house, dressed like the first, and then a third. There seemed to be an incredible number of these men – at one moment I thought I saw a dozen, then twenty. No, it was just the same three who kept appearing and disappearing as they came and went round their houses. It was six-thirty.

At this hour of day there was always a great commotion in Hans's house. The doors were flung open and banged shut. The six children in the house, who went to bed as late as their parents, got up before them. Their loud voices were as good as an alarm clock. From the sitting room the little ones shouted, '*Anaana*, where are my *kamiks*, my anorak, my – ?' '*Tassa!*' Hans shouted for silence. But the children clamoured for their mother now, and Hans's shouts had no effect.

Then Paulina came downstairs, wearing a faded floral dress, loose around the waist. Soon she was followed by Hans, barefooted, in long underwear, yellowed and dirty, the elastic loose around the ankles.

After their wash, Paulina called me. While I stood stripped to the waist in front of a bowl of water, scrubbing myself with a dirty, wet towel (I had clean towels in my luggage, but Hans insisted I use the one the whole family had used), the children gathered at the kitchen door to watch. There was a barrage of comments, and their parents started laughing. Paulina bustled between the kitchen and the sitting room, setting the table for breakfast. A good strong smell of coffee filled the house. Back in my room, I heard Paulina shouting, 'Mikili, *kaffemik!*'

Slices of bread and butter, jam and gruyère cheese were all laid out in generous quantities. The bread came ready-sliced from the *pisiniarfik* (general store): they spread the butter thickly, then a layer of jam or gooseberry jelly, and on top of all that a good slice of gruyère. A Danish delicacy.

I took a seat between Faré, the second youngest of the family, and Naja, the youngest girl. The four other children were already looking at me with a little less wonder in their faces. It had taken them less than twenty-four hours to get used to me. As Naja was making room for me, she even called me *qattann-gutiga*, my brother. Hans and Paulina sat at either end of the table. Suddenly Naja started crying. She wanted tea, not the usual coffee, but they were out of tea. Would her parents scold her to keep her quiet? Not at all! Leaving his cup of coffee unfinished, Hans went out to buy tea, then Paulina got up and made it. Naja's cup of milky coffee was poured into the kitchen sink, then she was served her tea – and all this was done gently, with tender words. But the little girl cried louder than ever, and now she was screaming. '*Sunaana, paninnguaq?*' (What's wrong now, my darling little daughter?), Paulina said, trying to comfort her. Naja didn't want to drink her tea out of her cup, but out of her saucer! The cup was removed without comment and the saucer placed in front of her. 'Do exactly as you please,' the mother said calmly, without a trace of exasperation. Naja poured the tea into the saucer, stuck out her lips, and slurped up the tea, to the vast amusement of the whole family. Meanwhile, Faré, seated on my right, put the big toe of his left foot in his mouth, with his other foot resting on the table almost in his jam.

Our breakfast ended with this scene, worthy of a Brueghel.

Paulina cleared the table. I helped her do the dishes while Hans played with Naja, hiding objects which she always triumphantly unearthed. This went on for a good while, with much laughter, when Paulina suddenly interrupted their game.

'*Tassa!*' she shouted. 'You're bothering the *Qallunaaq!*'*

'No, no, I like children.'

'How many *meeqqat* have you in your country?' she went on, holding up her fingers.

'None.'

'What? A big man like you? Don't you have a wife? Oh, you must find one here!'

'All right. But who'll take care of the children when I go back to Africa?'

'My husband will,' she said, as if it were the most natural thing in the world.

Here I must explain that of Paulina's six children the first two, Hendrik and Tage (pronounced 'Taë', the g being mute in Danish if between two vowels), had a Danish father, a workman now back on the mainland. The third, Assa, was by a Greenlander. The others, two little girls and a boy (Nuka,

---

* *Qallunaaq*: this term, applied to me by many other Greenlanders, does not mean 'white man' as some translate it. In its original sense it doesn't even mean 'foreigner'. One Greenlander translated it for me as 'he who comes from the south', but some of his compatriots disagreed. The correct meaning of the word has apparently disappeared because of an incorrect translation. Bear in mind that in Eskimo white is *qaqortoq*, black *qernertoq*, red *aapalaartoq*, etc. These terms are used to indicate the colour of an object, a landscape or a dog's coat, but are rarely used for a man.

Naja – meaning 'seagull' – and Faré), were by Hans, who was happy to have all these children under his roof.

It was eight o'clock. Hendrik, the eldest, put on his wellington boots and ran out to the stream, soon followed by all the others. When they left the house, the children just pushed open the front door, which had not even been locked for the night. The key had been lost long ago, and the lock would soon go the same way, for it was held on by a single screw in the upper left corner, and swung like a pendulum every time someone banged the door. The children slammed it so often and so hard that the whole house shook, and set the pendulum jiggling crazily. From the window, Paulina watched her children.

'With six *meeqqat,*' she sighed, 'I'm already an old woman at twenty-eight.'

'*Naamik!*' I protested, for she still looked youthful.

A little later the front door opened, this time so discreetly that only a slight squeak made us turn our heads. It was Louisa, their neighbour, accompanied by her daughter Sophia. Louisa was wearing a blue skirt and blouse, and walked with her toes turned in like a crab.

'*Pulaarpunga!*' (I've come to pay a visit!)

At eight o'clock in the morning! Timidly, she sat on the sofa, the nearest seat available. Her daughter snuggled against her. Paulina turned to the mother and daughter.

'*Kaffemik?*'

'*Aap.*'

'*Qaagit!*' (Come and sit at the table!) said Paulina, very free and easy.

Louisa accepted the invitation. Soon afterwards, the door opened again: another visitor. Then a fourth and a fifth, and so on. The visits had begun and would continue without interruption all day long. People were taking turns, it seemed. Those who had already had coffee at home drank another cup here when it was offered, and would drink more cups at other houses later. Each time new visitors arrived, Paulina put fresh cups on the table and also served me, herself and Hans. As I wanted to find out the local people's coffee capacity, I never refused a fresh cup, even though I had long ago drunk my fill.

'*Illit pikkori,*' (You're doing fine) Hans told me, and sent for *immiaq.*

The conversation grew livelier. Paulina fetched glasses. The words *baja* (beer), *akvavit* – a Danish spirit with an apple base – and *pisiniarfik* (the general store) kept cropping up in the conversation. Hendrik was sent off to the *pisiniarfik.*

At ten o'clock Adam came in.

'*Kaffemik?*'

'*Aap.*'

He sat down and turned to me:

'We'll be going out soon, Michel.'

'Where?'

'Visiting.'

He was wearing a grey suit and tie, while I was in my pullover.

'Wait till I get dressed.'

'No, no, you're fine like that,' Hans and Paulina chorused, getting ready to come with us.

Before leaving, Paulina made a fresh pot of coffee, which she left on the stove, and set cups and saucers on the table for any visitors who might arrive (since the door stayed unlocked) during our absence.

Then we, too, set out on a round of visits, starting with Adam's parents. His mother, overjoyed at seeing her son after twelve years' separation, had spent the previous evening brewing a barrel of *immiaq.* Hair flying, tipsy, and panting noisily, she greeted me by kissing my hand.

'*Aap*, I'm telling you!' she said, index finger solemnly raised. 'You have no family here, so I shall be a mother to you!'

Fresh cups of coffee from my adoptive mother, fresh glasses of *immiaq*.

We left a little later, surrounded and pursued by children. We came to the store and went in. It was a real supermarket! Under neon lights which shone inside despite the perpetual daylight, hundreds of tins were lined up on the shelves. There were bolts of cloth, boxes of matches, alarm clocks, pots and pans, guns. There was tobacco, coffee, tea; cans of beer, bags of potatoes, wheat flour, tins of milk and boxes of powdered milk, oranges, apples, and even bunches of bananas! There were woollen suits side by side with canvas anoraks and leather boots. Every kind of alcoholic drink was on display: whisky, gin, rum, vermouth, Cinzano, Dubonnet, cognac, *Kalhua*, vodka, *akvavit*. You could find everything at the *pisiniarfik*.

In the village, people kept greeting one another: 'Why! I was just on my way to your house!' 'Eeh! Go on and wait for me there, I'll be home soon. Yes, *massakkut*, see you soon!' I consulted my pocket diary to make sure this really was a Monday, a working day. Yet most of those we met were already tanked up, or coming from the *pisiniarfik* with one or two big cases of Tuborg or Carlsberg beer, which they usually carried on their hips. Some of them even invited me to a *bajamik*, a beer party, but Hans, who tended to regard me as his own property, absolutely refused: '*Naamik! Una ajorpoq!*' (No, he's not worth bothering about!) he repeated with every invitation I received. The number of those not worth visiting increased with every step we took. Eskimo custom requires a foreigner or a stranger to pay everyone a visit in the village, but we were skipping a great many houses. Hans was standing on his dignity, proud and haughty. Had he not stood at his door the night before, rigorously weeding out those who wanted to come and see this new kind of *Qallunaaq*? The poor unfortunates who were refused this honour (Hans was probably settling old scores) loitered round the house – so near and yet so far – looking up and pointing at me every time they glimpsed me through the window.

Hans was being warmly greeted from all sides. He had moved up in the world, and he knew it – it gave his pride a boost. After all, thanks to his wife, he was the official host of the first African to visit the country. That was bound to command respect.

Strolling along the bank of the fjord, we came to two kayaks, one of them covered with sealskin, the other with canvas painted white. They told me that these were the last two kayaks in the town. In other words, there were only two hunters left in Qaqortoq. Motorboats belonging to the inhabitants bobbed on the end of mooring ropes made fast to bollards or outcrops of rock.

We passed the landing-stage, the warehouse, and the little naval dockyard on our right. A curving street led to the *Gaest-ehjem*, a little hotel or guest house. Then the town petered out by the soccer field. Further on there were only mountains and deep valleys full of huge greenish stones, velvety soft with the mosses and lichens of the tundra.

In all, the village consisted of about eight kilometres of winding streets. The longest, which snaked along from the naval dockyard to the mountains behind Hans's house, covered at most one kilometre. Some Danes flaunted their status by using their cars to cover these short distances! Even taxis had been introduced into the village. The inhabitants would pile joyfully into them and go for a jaunt no further than the distance between two Métro stations in Paris.

Unable to go further, we walked back, taking a street that ran behind the *Forsamlinghus*, or town hall. We were supposed to be out paying visits, but because of Hans's standoffish attitude, our outing had turned into a sort of triumphal parade through the village. We would have returned home without having paid a single visit if we hadn't passed near a building where a group of old people standing on an upstairs veranda started waving to us. We waved back, and they sent a little boy who came running up to us to ask Hans if his 'foreigner' could pay them a visit.

'When?' Hans asked.

'*Massakkut,*' the boy replied.

'What is that building?' I asked.

'*Utoqqaat illuat!*' replied Hans, laughing merrily.

'Old folks' home,' Adam translated.

'Let's go!' I said.

Hans graciously agreed, and we made our way towards the old folks' home for Qaqortoq and the neighbouring villages.

Standing among them on the veranda, I was struck most of all by their small size, smaller than that of the other Greenlanders I had seen so far. They were not at all plump and chubby, but thin and tiny as Chinese. Some wore *kamiks*, others wellington boots. They wore thick black trousers tucked into the tops of their boots and long pullovers. Their hair, still quite black for the most part, was thick, straight and long, but cut so as to show a bit of the ears like big black berets pulled three quarters of the way down their foreheads, above slanted, laughing eyes.

They crowded round me and shook my hand. The old ladies uttered little gasps at my height. The men seemed intimidated as they looked up at me. Most of them were between sixty and sixty-five years old. Many had come to the home at sixty, the age of admission for men; for women it was fifty-five.

After the cups of coffee always served to visitors – instead of the calabash of fresh water we offer in my country – the old people took me on an inspection tour: the dining room, big and airy, with a high ceiling; then the library (among the books bound in black cloth were Eskimo translations of *The Three Musketeers*, the *Odyssey, Davy Crockett* and *Ivanhoe*), and finally the kitchen, with all the most up-to-date appliances. The superintendent was a Danish woman, and the old people rarely ate seal. To 'build them up' the fare was often Danish.

Now each of them wanted to show me his personal keepsakes. So we did the rounds of all the rooms, and I was struck by how clean and tidy they were – after all, they hadn't been expecting a visit. The rooms were large: they lived two to a room, the men's rooms on one side of the corridor, the women's on the other. With each room I entered I was impressed by the neat arrangement of all the objects in it. The daylight shone on two tables holding the personal effects of the two occupants: each had his own pipes, tobacco, newspapers, magazines and pencils in good order. In the corner of the men's rooms you always saw a toolbox; in the women's rooms, chests brought from their villages containing *ulu*, crescent-shaped knives used for scraping skins, as well as needles and bodkins of all sizes, and dried seal sinew – all needed for the work they do to earn their pocket money. The old men carve statuettes from walrus tusks, which they sell to the Danish officials, and the old ladies hand-sew elegant traditional sealskin costumes for the young people of the village, who no longer know how to make them. For thread they use the dried seal sinews, which are rolled against the cheek after being divided into fibres. One old woman gave me a demonstration. Two others shut themselves up for a while in their room, then emerged clothed in the resplendent national costume. It was total enchantment.

This women's costume from the southwest coast, a real masterpiece of patient handicraft, is composed of three pieces: the *kamiks*, the *takisut* and the *anorak*.

The women's *anorak* is distinguished from the men's by not having a hood. It is a sort of cloth tunic, chequered with dark and light squares like a chessboard, and decorated on the outside by the *nui*, a broad band covering the shoulders, chest and back, made entirely of multicoloured pearls. Subtle colour contrasts produce motifs with superb regular patterns. These motifs differ not just from one region to the next but even within the same village, sometimes within the same family. They are invented by the craftswoman, who develops her own decorative effects. In the old days, before there were any contacts with Europeans, the tinted bones of small fish were used instead of pearls to create the marvellous *nui*, this delightful feminine garment. The collar and cuffs of the *anorak* are bordered with an *ilupaaqusit*, a band of black skin.

The *kamiks*, made of treated sealskin with the hairs scraped off, are pure white. While the men's *kamiks*, which are black, stop at the knees, those of the women, decorated lower down with mosaics of little bits of coloured leather and then with lace around the knees, rise to the top of the thighs, where they too end in a horizontal band of black fur. Between these extraordinarily elegant boots and the tunic, there is the third and last item in this ensemble, the *takisut*: breeches made of sealskin, so short that they barely reach the tops of the boots, while the waist comes below the small of the back, where it is only just covered by the bottom of the tunic. So that, as an eminent Danish ethnologist notes, when you first see a Greenland woman in national costume bending down, you have the impression of an impending catastrophe.

The front of each thigh of the *takisut* is decorated with three vertical fur stripes, a dark one between two white ones.

Though the motifs decorating the *nui* have no particular significance, the colour of the *kamiks* is full of meaning. Small children of either sex wear red *kamiks*, the colour of the sun and of life itself. Differentiation appears around the age of seven, sometimes earlier, when little boys adopt the colour of the men's *kamiks*, and little girls that of the women's, together with the whole feminine costume, for it is assumed that at that age they know as much about life as their mothers . . . Thus there is no difference between the costume of a girl of fourteen after confirmation and that of a married woman, so that an outsider may well confuse them without running the risk of rebuke. But after the age of fifty, whether they are widows or not, the women change their beautiful white *kamiks* for black ones, still just as long, but with no lace at the knee. Sometimes they decorate them with modest and very sober leather mosaics. Their tunic remains the same, but it lacks the elegant collar of pearls and so is a little sad in appearance. Any woman who infringes this rule and dresses younger than her age invites the mocking little ditties that Greenlanders are so clever at improvising.

The word *anorak* (*annoraaq* in Greenlandic) comes from the Eskimo *anori*, the wind. This coat's main purpose is to protect the wearer from blizzards and cold. A windbreaker. But while the ones the women wear serve this practical function, they are also meant to make them beautiful. To make a complete female outfit requires five to six months' work, and the older women devote a good part of their time and ingenuity to such skilled tasks.

One of the women 'modelling' the costume for me was called Arnannguaq, 'the pretty little woman'.

'If you like,' she told me, 'I'll make you a fine white *anorak* for Sundays.'

'Thank you.'

A Greenland woman puts the very best of herself into making clothes to be worn by her husband or by some well-known person in the community. But while it is fashioned with the greatest care, a man's *anorak* is finished very quickly in comparison with a woman's. I asked Arnannguaq:

'If I bring the cloth tomorrow, when will my *anorak* be ready?'

She gave me a sly look and smiled.

'Oh, you know, a man's *anorak* is ready in two days.'

Amid the gales of laughter unleashed by these words, we returned to the dining room for a *kaffemik* (coffee).

Those old people seemed happy enough in their comfortable surroundings, but the tragedy of loneliness in old age was still with them: there was no family life in the Home. Contrary to what one might expect, not all the old people in the region lived there. Why, then, had these twenty or so left their families? I might still have believed it was because of the well-known custom by which the young used to kill off the old, considering them as useless mouths to feed, had I not later seen other old people in Qaqortoq enjoying a peaceful old age in the midst of their families. Nevertheless, the authority they hold in the village and at home is far exceeded by the power of the old in Africa, where as long as the grandfather is still alive, the father and even (in Togo) the aunt carry little weight in the family councils. In the village our elders, who are the masters of ceremony, will suddenly decide to hold a council meeting and summon all their children from the town, just like a king requesting the presence of his courtiers. And they arrange, accept, or reject the marriages of family members as they see fit. On the other hand, an old Greenlander will seldom contradict one of his own family. Here, the father neither scolds nor punishes a child, whereas in Togo he derives such authority from the respect due to age – a passive, abject, uncomplaining submission – and from the deep-seated belief that the old wield secret powers for good or ill, that not only families but even political regimes depend upon them and on the traditional chiefdoms. I cannot imagine our Togolese patriarchs ever agreeing to end their reign in old people's homes.

In the old days, both in Greenland and among the other Eskimos, the old people, so as not to encumber a migration, would elect to remain behind and die slowly in the abandoned igloos. It was a spontaneous, stoic, unforced decision, and one which to them seemed noble.

Today the old sometimes commit suicide. An old man may be driven to such an extremity when he is *kamappoq*, angry. Angry with himself. He goes out and never comes back. This happens particularly in winter: he leaves the house and walks a long way out on the frozen sea without heed for the places where the ice is soft, then all of a sudden – just as he had hoped – he sinks and is swallowed up. Sometimes he tells his family, and they do nothing to stop him. The old man has made up his mind and will not budge! Those who kill themselves in this way have often been great hunters. Diminished by old age and feeling themselves a burden to everyone, they don't take easily to their changed condition.

Aalu (Greenlandic form for Aaron), a former hunter now living in the home, started to tell a story, an old hunting story. Carried away, he talked so fast that Adam couldn't keep up in his translation. The cadenced phrases poured out, and after each flood of words, the listeners nodded their heads and murmured: '*Suuuu . . .!*' They had probably heard the same tale a hundred times, but they were all hanging on the storyteller's words. I had lost track of the narrative, so I watched the old man's movements. Seated on the floor in the middle of the room, his face shining with sweat, he was paddling his kayak, steering in and out of rocks. Suddenly his bright eyes focused on something ahead. He stared at it intently, removed his mitts, and rubbed his hands together. '*Issippoq!*' (How cold it is!)

Now we see him take something out of his kayak: the harpoon. He mimes each movement so well that I find myself holding my breath. One muscular arm launches the long harpoon, fatal to the seal, straight at the imaginary object that Aalu's eyes have never left for a second. Then he falls over on the floor. He now mimes the death throes of the *puisi* (seal) struck in the back of the neck, plunging underwater and dragging with it the sharp harpoon point. This detachable head, tied to a long leather thong, remains embedded in the animal's flesh, while the shaft floats to the surface. Attached to the other end of the thong, a bladder acting as a buoy enables the hunter to locate the wounded animal, which is diving deep. But the *puisi* is losing a lot of blood. '*Imaq aapilappoq!*' (The sea is all red!) He rises to the surface, and Aalu finishes him off. Now he is *togu*, dead. With the help of a tube (which he mimes with his closed fist) Aalu vigorously blows air into the animal's body through the hole of the wound. This procedure alone lasts for several minutes and provokes much appreciative laughter. The inflated seal is now floating. Aalu ties him up alongside his lightweight craft and paddles triumphantly back to us from his successful hunt. Now he's dragging the seal across the floor of the dining room, at the end of a rope through the animal's nose.

'That's a really big seal!' says Arnannguaq, getting up.

She cuts up the seal and shows us the skin.

'This will make a good *anorak* for Mikilissuaq!' she cries, as she mimes the skinning of the seal.

Never before had I seen a mime which could equal, and even outdo, the power of words.

We said good-bye to those fine old men and women in the Home, who begged me to come back and visit them often.

The next evening Adam and one of his cousins, Gerhart, came to take me out visiting. Once outside, we took the shortcut across the stream, jumping from one big stone to another. Another five minutes' walk and we were entering a modern building opposite the hospital. This was the residence of the Greenland student nurses, all of them girls.

On either side of the corridor were five rooms, each containing two bunk beds. Gerhart, who seemed well acquainted with the place, marched boldly through these rooms to find the girls, while Adam and I sat down in the hall. We didn't have long to wait: out came the girls, followed by our friend, all in high spirits, carrying a record player and records.

In readiness for tomorrow night's dance at the *Forsamlinghus*, Léa had her hair in curlers and had trouble turning her head. The hall was transformed into a dance floor. Beer started flowing in torrents, and we had to keep going to piss in the showers, turning the taps on after each visit.

'We're going to spend the night here,' Gerhart whispered to me. 'Twenty girls and not a man in the place, *ajunngilaq!* You see? But watch out – there's a woman superintendent, *Danski*, a Dane!'

Towards nine o'clock the music and happy voices fell silent, as if by magic, at the sight of an old Danish lady. This was the fearsome superintendent who, every evening at the same time, closed and locked the front door and took the key with her, after first inspecting all the rooms.

As soon as they heard her coming, the girls quickly hid us in their rooms. That is how Adam and I found ourselves in the bedroom of Lydia and Kathrina and compelled to spend the night there, Adam in the upper bunk with Kathrina, and I in the lower one with Lydia.

This little escapade kept us laughing all week.

The next day, towards evening, Gerhart called again, all smiles. He was wearing a clean shirt and new patent-leather shoes.

'There's a dance tonight! Have you forgotten?'

Somebody lent me a tie, and we went racing over the rocky

ground towards the *Forsamlinghus*. Every path was full of young men in smart cloth trousers and girls with long black hair tumbling down over red, European-made anoraks, all of them heading towards the village hall. We were hurrying because about ten o'clock the dance hall doors are shut and bolted to control the crowds; if you're late, it's just too bad! Gerhart, who had been there since nine o'clock, had noticed my absence and had come to fetch me before it was too late.

We arrived out of breath and made our way to the seats reserved for the men. In this hall, lit by electric light, the women were on seats lined up along one wall, and the men along the wall opposite. This was the general rule, Gerhart informed me; even the married men observed it. It seemed strange to me. Separated in this way, the two sexes looked each other over and exchanged glances, some inviting, some jealous. As soon as the music started, each man, his mind made up, hurried across to the women's side. It was a real stampede, a horde of black *kamiks* charging across the floor. In the commotion men shoved and elbowed each other without apology. When the brunt of the attack was over, the more timid youngsters were stuck with the wallflowers and the old ladies. The army of black *kamiks* had passed by.

Sometimes two men pounce upon the same girl, each one grabbing an arm. 'I was first!' you hear them both yell. Then it's not unusual to see a third man, cleverer than the first two, lead the young lady off into the dance while the other two gape with astonishment and then start a cheerful conversation, amazed to find that they share the same tastes.

That night the jukebox in the corner had been replaced by an all-male band consisting of an accordion player, two guitarists, and a hip-swinging vocalist who sang in English. Their long black hair, sometimes frantically dishevelled, naturally made them look like the Beatles, whom they were trying to imitate.

Now the dancers were swaying to a slow rhythm. The light shone down on a mixture of blond heads – the Danes – and dark heads – the Greenlanders. The Danish boys bent over their young partners, their big hands clasping them firmly round the waist.

For the next number, a twist, no one got up.

'You must be a good dancer,' Gerhart told me. 'You're making the others shy. No one wants to go out on the floor unless you do. Lydia *aana! Takkuk! Asasannguaq!*' (There she is, Lydia, your little sweetheart – look!) he added, giving me a wink.

So I crossed the room, all eyes on me, and led her out on the floor. Her beaming smile betrayed her feelings. We were the only dancers.

Whoever told me Eskimo girls were heavily built? Under Lydia's silk dress a graceful young body moved obediently. She followed my steps in perfect harmony. I held her close and caressed her neck, to wild applause. In the cheerful noise that broke out at the end of this tango, a young Greenlander in an overcoat strode across the floor, tapped me on the shoulder, and said: '*Kammassuak!*' (My good friend!) He slipped his hand inside his coat, brought out a half-full bottle of *akvavit*, and made me take a long swig.

The consumption of alcohol was forbidden here, and only lemonade and beer were on sale, which is why people did their drinking at home before leaving, and arrived at the dance maybe not drunk, but mellow. This got them all worked up and tended to bring out wild behaviour, even though generally they were shy!

Outside, the men were pissing against the walls.

The dance ended at about one o'clock in the morning. Outside we were back again in the cold, in the calm and gloom of nature, with sunlight, stern grey mountains, the white refraction from the sea of ice, and the sad-looking houses scattered among the rocks, each one long since known and numbered. I went back to Hans's house with Lydia.

At three in the morning, while we were sleeping, somebody came into the bedroom. He touched my hair and laughed. Awakened by the touch of a rough hand, I saw bending over me a bearded man with long black hair, who looked from me to Lydia and back again (bright sunlight was streaming through the window), and kept repeating with a blissful smile: '*Kusanaq! Kusanaq!*' (Pretty! Pretty!) I watched him nervously. Was he drunk? Frightened by the sight of him, Lydia huddled closer and closer. Twice he went into Hans and Paulina's room, walked down to the kitchen, helped himself to some *ammassat*, then left.

Of course the house was never locked at night. Next morning, when I asked the owners who it was, they didn't know, but they didn't seem too worried.

In the evening I went to see Lydia, who was in bed with a temperature and couldn't come to the cinema with me. So I went to find Gerhart.

'Lydia can't come? Never mind, you can take Christine. If she's not at the cinema, we'll see her later at the Nanoq.'

I knew that the Nanoq was the other dance hall, open only on Sundays, when the town hall was used as a cinema. But who was Christine?

'She works at the *pisiniarfik*. You'll see, she's all right, *kusanaq!* All the *Danski* are after her.'

At the *Forsamlinghus* a gigantic poster advertised *The Man in the Raincoat*, a French movie with Eddie Constantine. I had already seen it, but now that the *Martin S* had left, this old film represented my only remaining link with the European world. So at about six I went with Gerhart to the 'cinema'. The hall was packed.

The film had Danish subtitles. Ten minutes after it started, the picture suddenly froze on the screen.

'Has the projector broken down?' I asked Gerhart.

'*Naamik!*' he answered, settling back in his seat.

A muffled voice emerged from a loudspeaker to explain in Eskimo the part of the movie we'd just been watching. Then the picture started up again, and the same odd procedure was repeated every ten minutes, which made the show very long, so that they had to speed up, or even cut, certain scenes.

The Danes can't stand dubbing because it seems incongruous to them that a foreign actor should address them in Danish (and here they're not far wrong, since Danish is not a language of beautiful sounds), so all foreign films are screened in the original version with Danish subtitles. But this is where things get complicated: they have to show these foreign films with Danish subtitles to Greenland Eskimos, who rarely understand Danish, let alone the foreign language spoken in the film. Adding Eskimo subtitles would be expensive, so the solution is to stop the picture every ten minutes and, with the hall still in darkness, explain to the audience what has just been happening. As you can imagine, this was no fun for someone who spoke French!

On our way out, I asked Gerhart if he had understood the movie.

'*Aap!*' he replied with a broad smile.

And that was that.

A few days later, Gerhart and I went to visit Lydia.

'*Asavakkit!*' (I love you!) she cried when she greeted me. Deeply moved, I stroked her cheek. As we were leaving, she told me:

'Come back with Adam this evening at seven.'

'Why with Adam?'

'So that we can spend the night with Kathrina like the last time,' she explained.

That evening we refused all other invitations. When we got to Lydia's room a little earlier than expected, we saw Karl, Adam's brother, lying naked in the bed beside Lydia! They were drinking beer and laughing. Seeing them lying there side by side, I couldn't help feeling upset. Karl pulled Lydia against him and beckoned to me to slip into bed beside them. I refused and left without saying a word, though not before having to endure a tirade by Adam, who thundered against the Danes for not having 'improved the material living standards of my people and modernized the country'. He preferred the Americans. Lydia disagreed with him. Karl fell asleep with his head on Lydia's breast. All for Karl at the moment, she had no eyes for me.

I went to visit one of her neighbours, Anita, and when I came back Lydia was alone. Karl and Adam had gone. Now it was my turn to make a speech – not against material living standards but promiscuity. I spoke in a mixture of Danish and Eskimo. 'How can you say you love me? What do people in this country mean by *asavakkit?*' But nothing I said made the slightest impact on Lydia, who seemed not to understand my attitude. So then I played my big scene and accused her of being unfaithful. In the end, she started crying. As I stood there in the middle of her room, her tears left me cold. Kathrina arrived and didn't say a word when she saw her friend crying into the pillow, but when she learned what was happening she accused me of being jealous. To my astonishment, she didn't understand why I was angry either. In Greenland, jealousy is frowned upon. Kathrina explained:

'Before you came, Karl and Lydia were always together.'

'Is that so?'

I left the room with plenty of food for thought. So Karl was Lydia's boyfriend. In that case, who should be jealous? Certainly it ought to be Karl!

I went back to Adam's, and the first person I saw was Karl.

He called me his brother, sat me down, and brought me a cup of coffee. And that wasn't the half of it. Adam, a Europeanized Greenlander, asked me as he sipped his coffee:

'Have you given Kathrina a try? She's nice.'

'But she's *your* girlfriend!' I answered, baffled.

'Oh, yes, but . . .'

I understood and looked away.

This sharing of bedmates among friends was unquestionably tempting, and I was quite willing to share other men's girls – but not my own. My behaviour at this time, and my own way of seeing things, must have been a fair reflection of our two worlds, theirs and mine, in this respect irreconcilable.

Greenland morality was beginning to disgust me. One evening I went to see Lydia, who had just finished work. We went dancing and came back to my place to sleep. The next morning, the mother of Eric – Gerhart's brother – came into my room, bringing me some native candlesticks. She didn't knock before coming in, and winked and grinned at finding a girl in my bed. It was impossible to have any private life. Now that she had woken us up, we went downstairs for coffee. Then Karl arrived. Lydia ran to meet him! They sat down side by side at the other end of the table, and I could see Lydia's hand moving up Karl's thigh. Then almost immediately afterwards she came back and sat beside me. I left the house in a sulk.

Two women – each standing at the door of her house – invited me in, disputing my company between them. I couldn't make up my mind, so I decided to visit each one in turn.

The first one: The two beds in the bedroom had clothes scattered helter-skelter all over them. There were photos on the walls – here some pictures of Jesus, there one of the Virgin holding the Holy Infant in her arms – and a coloured map of Greenland, a schoolboy's work.

'My son drew it. I have two *meeqqat*, a daughter and a son, both with a Danish father.'

She bent over me to kiss me. I avoided her lips. She nestled close to me, calling me 'darling.' A crackling radio played music. Then she tried to sell me some pearl necklaces, spreading them out on the table.

'I have no money,' I told her.

Ignoring me completely after that, she began to iron a skirt with a pattern of little red flowers.

'It's for tonight. Are you coming to the dance?'

'Perhaps.'

'Do come – you must!'

The iron, made of real iron and practically an antique, was heated on the big cooking range which also served as a radiator. In one of the beds was a doll.

The second one: She was drunk, lying on her bed staring at the ceiling. Although she was awake she hadn't heard me knock at the door, but I went in anyhow, stood beside her and spoke to her. For some moments she still did not move, her eyes still wide open, fixed on the ceiling. Finally she roused herself, sat up in bed (here there were two beds, too), and looked around without seeing me: it was like the wandering stare of the cross-eyed. I refused to sit beside her.

'Come again,' she said in English, without really knowing what she was saying.

I was feeling more and more disillusioned. Dancing and drinking were all they had in life. This was not the Greenland of my dreams. I wanted to live with seal hunters, ride in a sledge, sleep in an igloo! But apart from two kayaks, there were no seal hunters left in Qaqortoq, not a single sledge, not a husky. And not one single igloo!

To find out what remained of the traditional pursuits, I consulted some statistics and learned that the population of

Qaqortoq, in the year 1959–60, had killed 1 bear, 1 porpoise, 413 seals, and 61 foxes – a total of 476 animals in one year for a community then numbering 1,741 inhabitants. Those figures had fallen every year since then.

'Nowadays,' I was told, 'the population lives almost entirely on catching shrimp and cod.'

But apart from Eric, a man called Locarno – a Swiss who had taken Danish citizenship – and one or two others who often left the village on fishing expeditions, none of the people I met seemed to have any definite job. Gerhart trailed me around with him day and night and seemed to do nothing. What about Karl? He gave me the impression of being a parasite living off his brothers, especially Adam. As for Hans, who supposedly worked at the naval dockyard, not once since my arrival had he gone to work. Yet Paulina was always offering coffee and drinks to visitors. How did she get the money? Well, let's face it: a lot of able-bodied Greenlanders simply live on allowances from the Danish government.

Why is this? Children are sent to school but are not taught anything about the traditional activities. Even worse, that way of life is disparaged to their faces, although it is their own. When they grow up, they can't even paddle a kayak. That's how things are for the Greenlanders on the southern coast.

'But are there still places with seal hunters and huskies, sledges and kayaks?' I asked.

'*Avannamuut!*' (You must go further north!)

'Where exactly?'

'*Naluvara!* (How should I know!) *Immaqa* (perhaps) Sisimiut, Ilulissat, Tûli . . .'

The names of these three communities in Danish are Holsteinsborg, Jakobshavn and Thule, all distant places.

'That's where you'll see the Eskimos!'

And these civilized southern Greenlanders, descendants of

Eskimos, now spoke the word 'Eskimo' with contempt, as the Algonquians used to!

All the same, I made ready for a trip to the Far North, even planning to spend several weeks in each of the villages on my way.

Local transport posed a problem. A journey from Copenhagen to any large community in Greenland is easier than a trip from one locality to the next. There are a few helicopters, but flights depend on atmospheric conditions. You can wait day after day for a whole month before taking off!

'You can sail from here or Narsaq on the *Kununnguaq*, which goes up to Upernavik, near Thule, every fortnight from June to September.'

So it was possible to make that long haul. But my plan to live for some time in each village on the way didn't fit in with this scheme. Hans and Gerhart talked it over and told me that if I travelled as I was planning to do, I would be completely iced in for six months in the region of Egedesminde (Aasiaat), where I would find myself having to wait until the following summer before continuing northward by ship. (This is what actually happened.)

'Or else, if the winter catches you at Sisimiut (Holsteinsborg), you could possibly continue your journey by dog sled.'

So I had to make very detailed plans, which I couldn't do without knowing the itineraries of the smaller coastal ships. And that required further information that no one had been able to provide in Copenhagen. I was all the more desperately determined on this scheme, because the idle life in Qaqortoq was getting on my nerves. I needed a direction and open spaces!

But it was not until August 12 (after a long stay in hospital with a suspected venereal disease – or some tropical illness) that I was finally able to leave.

At half-past seven that evening, I said farewell to Lydia and my hosts and made my way to the harbour. Adam came with me, and he and a young boy helped carry my luggage on board the *Kununnguaq*.

We sailed from Qaqortoq at 9.30 P.M. Even at that late hour, it is still daylight in the month of August. All the inhabitants were on the landing stage to watch the ship put out to sea. They wished me a good journey and a happy return. But I never went back to Qaqortoq, 'the White One'.

3

Fishing for Sea Wolf

All the time we were weaving our way out of the fjord, the sea was calm and flat as a river; on either side, mountain followed mountain. We passed small islands and icebergs. The sun rose at two in the morning behind the ship, in the direction of Qaqortoq, which I could no longer make out among the high grey mountains.

I was not travelling in a private cabin but with the Greenlanders, under the poop where we slept all together like sardines, on long benches used as seats during the day. Although there was one section for men and one for women, we always mingled.

At ten-thirty the next morning we arrived at Arsuk. This village is near Ivituut, a busier village where they mine cryolite, an aluminium ore extracted in the form of stones which are icy to the touch (hence its name). As there is no wharf at Arsuk, we dropped anchor offshore, and a motorboat plied between village and ship. We soon set off again, and at five in the evening arrived at Frederikshåb, where I disembarked.

Frederikshåb is three hundred kilometres from Qaqortoq and has a population of 1,040. Its name in Eskimo, Paamiut, means 'The Dwellers at the River Mouth'. The town is surrounded by islets and reefs blocking access to the harbour, so we landed as we had at Arsuk.

I was put up by the Steffensens, a Greenland family. The wife, after showing me into the sitting room, opened the bedroom door to introduce her husband. He was drunk and sound asleep. Casually, she switched on the light for me to see him better.

Their only son, Steffen, a crew-cut boy of fifteen, was sitting on the sofa. He helped me to stow my luggage, and then he took me on a tour of the village and a visit to the little dance hall. We went to bed around midnight, in the sitting room. Steffen's father was still asleep.

The drunken man woke up in the middle of the night and came into the sitting room, clad in an open-necked check shirt and black trousers, and holding his wife by the hand. The coal stove was still burning, and its red glow made their fat round faces shine as they came out of their room. The wife had put on a shapeless old dress, of the kind which most Greenland women wear as a nightgown. They both approached the stove, like trolls crouching in front of a forest bonfire, and then, as the wife stoked it for the night, the husband staggered to the front door, where there was a plastic bucket. There, almost under my nose, he unbuttoned his crumpled black trousers that glittered with bits of dried mucus. After emptying his bladder, he stumbled over my suitcases and nearly fell down. His wife now went to the bucket, lifted her frock (she wore nothing underneath), and squatted over the receptacle.

After that, she rearranged my bags and told her husband that they had taken in a stranger under their roof. She pointed at me and the pair of them, thinking me asleep, crept over, bent down, and examined me closely. 'He really is black!' said the man. Then they went back to bed. At dawn their son Steffen, also thinking I was asleep, sat up quietly and stared at me for a long time, even though he had spent the evening with me. Three times he stretched out his arm and gently stroked my hair, his eyes shining with indescribable curiosity.

So it was not until the next day, Saturday, that I really got to know my new hosts. Steffensen was fifty and his wife about ten years younger, though she looked older than he, being more wrinkled and withered, with very prominent cheekbones. They both came from Arsuk.

I opened my rucksack for a change of clothes and found it all wet and stinking of urine. The material was even wet inside.

'Did Steffensen have that much piss in him?' I asked, turning towards him.

The question came out so unexpectedly that we all started to laugh. I took out a sheet of paper with a large stain and held it up, saying:

'Look! It's as big as the map of Greenland!'

That gave us something to laugh about for the rest of the day.

All over the country in August the weather is rainy and foggy, and for those Greenlanders who have no other activity than what the Danes provide, the long weekend days are appallingly empty. At the windows of the neighbouring houses, weather-watching faces blank with boredom gazed through the cold panes.

Behind the Steffensens' house the long, widening valley stretched like an ancient glacial trough to the sea, where icebergs towered. The valley floor was relatively flat, and as the scattered houses stood about thirty metres apart, between them I could clearly see the wooden church in the centre of the village. The steep angles of its roof and its multiple belfries gave it the look of a Chinese carving. To the left, a line of telegraph poles ran straight across the plain.

Paamiut is divided into two equal parts by the main street, the Gerth Egedesvej or Gerth Egede Street, which is intersected by fourteen smaller streets, only four of which had acquired names as yet. There is a Waterworks Street (the Vandværksvej), an East Bridge Street (the Østerbrovej) – though there is no bridge – and a Cemetery Street. Not far from this last were eight oblong shacks covered with olive-green tarpaulins like an army camp and standing right out in the open, in a deep bend of the valley. They were inhabited by Greenlanders who had practically no contacts with the rest of the town, young people who had left surrounding villages to settle in Paamiut, in obedience to a government policy of concentrating the island's population in the bigger communities of the south, to make administration easier. These young men worked at the naval dockyard or, if they had been hunters, tried to become fishermen since shrimp and cod fishing are seen as the great hopes for Greenland's economic future.

The shacks that housed these newcomers were called *Diogenehytter*, huts of Diogenes – not that those young men were happy in the role! The most surprising thing was that their fellow Greenlanders didn't offer them hospitality to get them out of those freezing shacks, stuck between the cemetery and an arm of the fjord where rubbish was dumped – rotten wood, old beer and sugar cartons, sacks, crumpled or smashed containers, plastic bags, birds' wings, sheep heads, empty cans, old boots, glass and plastic bottles, worn-out gloves, women's stockings. Acrid smoke rose from the heaps of garbage and ravens flew overhead.

Near the eight tarpaulin shacks stood the poorest houses in the village, where happy voices were rarely heard, while at the other end of the village were two long buildings housing Danish labourers, who had a whale of a time every evening: loud music bellowed everywhere, and in the street you were deafened by the competing record players. Even in the Arctic, equality is only a word.

These Danish workmen are bachelors and live in Greenland with *kiffaqs*, local servant girls whom they hire in droves to do washing and make coffee. Their casual loves breed children to populate the south. Then one day the workers go back to Denmark, abandoning their *kiffaqs*, babies and all.

*

Since my arrival two days before, the children had spent all day hanging around the house where I was living. Once they knew my name and age, and above all once they had shaken my hand, they became my great friends, and would turn up unannounced in little groups to pay me a visit. They asked me where I would like to go and what I would like to see, took me off by boat into the neighbouring fjords, taught me their language, and passed on village gossip.

At about eight o'clock on Sunday evening they came to fetch me to go and pick the *paarnat*, little black berries which grow in the mountains. They took me along with them and made me sit down in a valley while they went gathering the tiny fruits which they brought back by the handful and cheerfully shared. The oldest of these young friends were ten years of age.

'*Africa nuanni?*' they asked.

'Yes, Africa is a beautiful country.'

'*Soorlu tamaani?*' (As beautiful as our country?)

Love of their country and of what is theirs comes first with them. For example, when offering food to a foreigner, they would often ask with the first bite, '*Mamarpok, ilaa?*' (It's good, isn't it?) Their joy was immense when you replied, '*Ab, mamarpok.*' (Yes, it's good.)

I told them that Africa is just as beautiful as Greenland. They made me tell them all about Africa. We talked about forests and wild beasts. They had heard of these at school but, never having seen a tree, their curiosity bubbled over. How did a tree grow? What height did it reach? I did my best to explain. While we were chatting like this, whole families gathered *paarnat* in the mountains, filling plastic buckets with the berries. We saw a girl, and one of my young friends indicated her with a wink and asked me:

'*Una, mamaq, ilaa?*' (She's good-looking, isn't she?)

'*Aap! Mamaq assut!*'

My answer made them jump for joy. They shouted with laughter and clapped their hands. The girl, who luckily for us had heard nothing, disappeared around the hillside, and we went on eating our berries.

A little later we started up a long, beautiful, grassy valley shaded by the great chain of mountains stretching to our left. At the other end, the late evening sun was reflected in a little lake. Brooklets ran murmuring between huge rocks. A thick moss carpet muffled our footsteps. I was walking in the midst of my young friends.

'How big he is!' one of them cried.

'He's my brother!' said another, grabbing me by the hand.

'No, he's mine!' cried a little girl, taking my other hand.

Then they invited me to visit their school. I went there the next day, a Monday, and was impressed by the rapport between the Danish teachers and their pupils, who talked together quite freely. To get to know their pupils and understand them better, the teachers learned Eskimo in the evenings. Teachers and pupils were friends, in a gentle atmosphere that did far more for these children than the severe discipline and frequent canings employed by our teachers in Africa. Never before had I seen children so happy to go to school (even in the Far North during the most intense cold) and so relaxed in class. My only criticism is that the teachers neglect native Eskimo culture and try to teach the children standards valid only in Europe.

At eleven o'clock in the morning of Saturday, August 28, I arrived at Godthåb, the capital, called in Eskimo Nuuk, or 'Hood'; the name is supposed to come from a mountain that overlooks the town – a lofty, saw-toothed peak whose highest point looks like a head peeping out of an anorak hood.

The town's Danish name, Godthåb (pronounced Gott-hoob), means 'Good Hope'. Indeed, lying in the Davis Strait, between the thick permanent icefield of the south and the dismal regions of the north, the peninsula housing Nuuk is well situated: ships from Europe can cut in from the open sea and often need only to steer through a few ice floes to reach port, so that the capital is accessible all year round.

The harbour, with its metal framework and massive beams supporting a reinforced concrete surface, is fairly big, with mooring for five or six cargo boats. To the right, an extension of this dock is reserved for fishing boats.

Some twenty people had gathered on the quay, and their number kept growing. A bus, the only one in Greenland, was waiting for the passengers.

Eleven shops and stores, a sausage-seller's van like those in the streets of Copenhagen, two churches, two hospitals, one of them the imposing 'Dronning Ingrids Sanatorium' (Queen Ingrid's Sanatorium), four 'café-bars', a radio station, a fire station, a soccer pitch, three kindergartens, two schools, one high school, a girls' boarding school, a municipal library, and a hotel, together with 644 houses, made up a town which then had nearly five thousand inhabitants. Here, in the capital, was Greenland's only prison, with six cells. It was as big as two ordinary houses combined, built on one level, and it would escape notice but for the wire fence surrounding it – an unusual sight in Greenland. My first impression was of a kind of chicken run as I walked around the four-metre fence. The gate, also covered with strong wire, was wide open. After crossing the yard and entering the building, you came to a short corridor. At the end, the warder sat at a table next to the kitchen.

'At present I've only five guests, all young men of twenty-five to thirty. Two of them come from the south, a third from here in Godthåb, and the other two from the north, one from Holsteinsborg and the other from Egedesminde.'

The jailer was unarmed.

'In two years' time we expect to have a prison with eighteen cells, because of the rise in crimes and misdemeanours during the past few years.'

'What are the most common types of crime?'

'Theft and murder.'

'May I visit your charges?'

'They're free during the day and aren't usually here. But you can visit the cells.'

Each cell contained an open cupboard, a table placed under the window, and a bed covered with very clean sheets and a quilt. The varnished body of a fine guitar gleamed at me from one of the beds; I leafed through two thick pornographic books full of photos which I found on a shelf over the bed. Each cell was arranged and decorated according to the personal tastes of the occupant. The last one on the right was unoccupied; through the open door I saw only a mattress and a bare table.

We were visiting the kitchen when a young prisoner of twenty-seven, a southerner, walked in. He was wearing trousers and a white shirt with no collar. He had been living in this jail for six months.

'Only one more week to go,' he told me.

'What's it like here? Not too bad, eh?'

A smile, a slight shrug.

The prisoners only had to stay in jail at night. The gate was locked at six in the evening and opened the next morning about half-past eight. Each prisoner worked wherever he wanted and paid for his prison meals out of his wages.

'I work in the kitchen at the hospital, where I make 2 kroner and 91 øre* an hour. Sometimes I work at the harbour helping

* Øre: unit of currency in Denmark and Scandinavian countries. One hundred øre make one kroner, or crown. Though the coinage is the same as in Denmark, the Greenland banknotes differ from those in Denmark. The one complaint is that there is no bank in Greenland.

to unload cargo when a ship arrives, to make a little more money. No one comes to supervise us when we go out to work.'

'So you can escape if you want to?'

'Oh, yes, I even tried it once.'

He gave a hearty laugh at the memory, and his dark eyes sparkled between their slanting lids as he threw a glance at the jailer, who was also laughing.

'It wasn't in the daytime, you know,' he went on, 'but at night. I climbed over the fence there' (he took me outside to show me the place, while the easy-going jailer went and sat down at his table again), 'and I managed to get over, just there. It's not hard, you see. I got scratched, but I managed to get out without making a noise. I wanted to take a boat back to my village, but they picked me up the next day while I was drinking at a friend's house. That's why I've got to do the full six months.'

Generally, prisoners who behave well and have regular work are released halfway through their sentences.

'If you're that free, why try to escape?'

'I thought the white men's law was unjust. They throw us in prison for the same stretch as they do for a crime in Europe, yet they keep telling us that our expectation of life here is shorter than theirs. A sentence of six months in Denmark ought to be three months for us.'

The young prisoner obviously had no idea what prison life is like in Europe.

'What crime did you commit?'

'I stole something.'

Minor offences are tried in Greenland, but murderers are taken by ship to Copenhagen, where they stand trial and are sentenced before being sent back and 'incarcerated' in Greenland.

'Do you wear uniforms?' I asked the young man, thinking of the barefoot prisoners in Togo, with their khaki uniforms – shorts and a grubby shirt with a big number sewn on front and back.

'No. We dress as we like.'

Outwardly, nothing distinguishes them from the other inhabitants, and that is psychologically important. Though a few people steer clear of them, most of the inhabitants don't, so the prisoners are free to go into people's houses and also into drinking places, though they're allowed to drink only coffee and *sodavand*, fruit juice. Alcohol is forbidden, but they drink it at their friends' houses –'just a drop', though, for a friend too lavish with his drink might get into trouble with the authorities. Finally, they're not allowed to bring girlfriends back at night, but they have plenty of free time during the day.

To get to the fishing port the next day, I ventured down a never-ending stairway set into the steep side of the cliff. The fishermen's houses clung to this rocky slope. Nets were drying on the roofs, and bunches of slaughtered seagulls hung from the doors. Once at the bottom, I walked onto a narrow bridge; fishermen sitting next to rusty anchors were mending nets slung from the single railing which ran along one side. I went on board a fishing boat belonging to a man called Christian, and asked if I could join him on a fishing expedition.

'It takes some time, you know. We expect to be out for a week, sometimes more.'

'That's all right.'

'Come back tomorrow, perhaps we might be sailing then.'

Immaqa! Perhaps! (The k is pronounced rather like the guttural French r.) That is the answer you nearly always get when you ask a Greenlander a direct question.

Immaqa! Ever since Sunday it had been 'perhaps'! Now it was Wednesday, and still we hadn't sailed. The next day, Thursday, I met Christian at about eleven o'clock. He was coming back from the store, laden with provisions – bread, sugar, coffee, a pot of mustard, and a can of gasoline. Obviously, we were going to sea at last. Buying gasoline was bound to be a good sign.

'Are we leaving tomorrow, Christian?'

'*Naamik!*'

'When, then? *Ullumi?*' (Today?)

'*Suuuu . . . Massakkut!*'

'And you didn't tell me!'

'I was waiting for you!'

I hurried back to the house to drop off my things, then returned to the harbour. We set out at half-past midday on board one of the smallish, two-masted ships so numerous in this country. They have a deckhouse like a sentry box, right in the middle of the bridge. Inside that is the steering gear – compass, rudder, ship's radiotelegraph, depth-sounder, and logline, this last an apparatus for measuring the vessel's speed. On the afterdeck, a hatch led down to a small cabin which would house all four of us for the whole expedition. Our only warmth came from an ordinary oil heater.

Transforming the Greenlanders from hunters into fishermen was an uphill struggle. From Greenland to Alaska, all the peoples known as Eskimos used to be divided into eaters of seal, caribou or fish. Those who ate meat did not eat fish and vice versa; those who ate both were few. Thus, a real crisis occurred in the Thule region when the seal hunting was particularly bad one year, and some inhabitants actually starved to death when there was fish for the taking.

Another incident, which occurred shortly before my arrival in Christianshåb, is worth noting. One day the men employed at the fish factory dropped everything and ran for their kayaks, because a herd of seals had been spotted in the bay. Their spur-of-the-moment hunt went well, and they never did go back to the factory.

The world of the hunters-turned-fishermen takes time to evolve. They don't become fish-eaters right away: they sell the fish to the factory, but while they are fishing they hope to hunt

A couple fishing at Narsaq

seal for their own use. And though the bridge of our boat was piled with nets and with tubs full of ropes, my companions also carried the equipment for killing and skinning seals. Which is why, one hour after our departure, we slipped into a cove, and Christian gave me the tiller while the three Greenlanders crouched in the bow, armed with rifles, scanning the surface of the tranquil water. I could see nothing.

'What are you looking for?'

'*Puisi!*'

The motor was only idling, but its vibrations scared off the seals, which are alert and suspicious. We sailed from one inlet to another. All afternoon there was talk only of seals, sea birds, ptarmigans, eiders, guillemots – everything but fish! Hours passed. My companions didn't care about the wasted fuel, though not a single seal was killed, or even a bird. Night began to fall, and they had to give up.

It was September. The sun, which hadn't set for months, was beginning to decline. At half-past four in the afternoon the sky was low and sombre. Only a gleam still gilded the horizon behind the peaks. We sailed past a mountain with an extended ridge which petered out into a string of reefs more than a kilometre in length, with a marker buoy at the far end. This long chain of reefs, which from a distance I had thought to be a single spur, turned out to be split into sections, with room to pass between. The area was strewn with sunken reefs.

'Where are we going to fish?'

'At Fiskenæsset.' (Qeqertarsuatsiaat)

'But we're putting in to Færingehavn.' (Kangerluarsoruseq)

'*Aap!* We'll go to the cinema tonight.'

Situated sixty kilometres south of the capital, Færingehavn, with a population of only thirteen, is an 'international port'. Foreign trawlers take on fuel there from huge oil tanks set at the foot of high, snow-capped peaks. We entered the harbour by a narrow channel between two mountains and found ourselves facing five houses – two red, two yellow, and a bigger white one, with two storeys and a lot of windows. In addition to these houses there was a seamen's hostel and, facing the harbour, a factory where they apparently converted fish or fish heads into powder. In a hamlet of thirteen inhabitants, including children, it didn't look as if this factory required an enormous work force to keep it going. The dock was fairly long, to accommodate trawlers, and went off at a slight angle at the end. We tied up at the other end, next to a fishing boat.

The one street of the hamlet ended in a wooden bridge leading to the seamen's hostel, which we entered. The place doubled as a church, so along with chairs and drinking tables, a cloakroom and a kitchen, there was a pulpit and two pianos, the newer one locked shut and standing in a corner. The older one was at the disposal of anyone who wanted to bang out a tune. Tattered hymnbooks overflowed a chest. There were shelves of newspapers, most of them Scandinavian. Those who didn't want to read or sit at the piano could play with amusement machines, like the bar football game, or stare at the paintings of Greenland views and scenes hanging on the walls. Or they could simply drink: there was a staff of Greenlanders to serve the sailors and fishermen.

We ordered coffee. Thirteen young Spanish seamen were kicking up a deafening din. They talked about Spain, which they had left three months before, and complained about the cold.

Two other young seamen, Finns, came reeling in drunk, and sat down like disjointed puppets. They ordered coffee but started to snore in each other's faces, heads propped on folded arms, while the coffee steamed on the table. The waitress woke them up, and they gazed round the room with haggard eyes, paid, and got up to leave on legs which could hardly carry them. The waitress took away their untouched coffee. It was here at Færingehavn last year that a young Finn, roaring drunk, cut off his penis and threw it in the sea. He was rushed to the dispensary but died soon after.

The friends we left behind at Nuuk must have imagined that by now we were busy fishing, but far from it. We spent part of a cheerful evening in the seamen's hostel and even came back a second time before going to see *Don Camillo*, starring Fernandel. Three flags decorated the little hall – the Danish to the left, the Norwegian to the right, and the Swedish in the middle above the screen.

Driving rain and a fierce wind met us as we left. When we reached our boat's little cabin, soaked through and freezing, we realized that the oil stove was leaking. The tank had a hole in it and air bubbles were escaping, but the stove still worked – it was just that we had to keep pumping to maintain the pressure, which caused fuel to squirt out and splash us in the face. We ate a meal of bread and butter and pâté, with cheese and strawberry jam, swallowing hot coffee from a bowl passed from hand to hand. Then we turned off the stove and blocked up the hole by daubing paint on it with a screwdriver. The cabin stank of reindeer and sealskins, burnt oil, and rancid sweat. The floor was foul. Spending the night there was torture. Christian made me sleep in his own usual place, on a sealskin in the recess down the left-hand side. This gave me a cozy but funereal feeling: the nook was like a coffin, man-sized and oblong in shape. I rolled up my overcoat under my head for a pillow. Abel slept in the recess opposite, and as there were only two of these, Christian and Jørgen (pronounced Yonn) lay head-to-toe on a bench underneath Abel's improvised bunk. The storm roared outside, and we could hear the lashing of rain against the hull. The boat was rolling heavily; finally, fearing she might heel over, we got up in the middle of the night and left the harbour for another mooring near a landing-stage. It was so pitch black during this maneuver that Abel stood in the bow and shone a torch on the mountain slopes to guide us. Another fishing vessel left harbour at the same time and slowly followed us, showing a red light to port and a green to starboard, with a white lamp at the top of one mast. We moored in a bay where there were several boats already, and all night long listened to a concert of creaks and groans.

There was still no question of fishing the next day, Friday, but now we had a good excuse – the storm at sea. The flimsy little vessels took a dreadful battering in the bay. All day long the sky stayed grey and the wind howled over the foaming surface of the sea. The men in the next ship were plucking eider ducks for boiling, and the gale literally tore the feathers from their hands. The thick whitish smoke from the factory, which stank of rotten fish, streamed from right to left like steam from a speeding locomotive; or else, suddenly beaten downwards by a violent gust of wind, it looked like a mass of foaming water falling from a great height.

In the evening a lull enabled us to go back to see a western. On the way to the cinema, a Dane caught sight of me and shouted, 'Hell, the poor black man – he'll be frozen stiff!' He offered me *akvavit*, and the two of us drank it from the bottle on the street.

When we came out of the cinema, the wind was blowing worse than during the day. We had to walk with our heads turned, sometimes bending, sometimes leaning into the wind, while the rain whipped at our bodies. Back in the boat, we changed moorings once again, making for a more sheltered bay.

On Sunday we set out from Færingehavn against wind and tide, and the phrase was never truer. The storm was still raging, and we had to cover eighty-eight kilometres of furious sea to reach Fiskenæsset. Avoiding the high seas, we steered through narrow channels in between beaconed islets. Yet we stopped in every inlet, hunting for the seals that were always on the fishermen's minds. Abel and Jørgen stood in the bow, rifles at the ready, but still no seals! They shot at birds without hitting a single one. I was beginning to wonder just how skilled my shipmates were.

A little later, something else cropped up to divert their attention. As we were emerging from a channel, we spotted a fishing vessel whose engine was less powerful than ours, making poor headway in the storm. She was so low in the water that she seemed close to foundering.

'Shall we take her in tow?' proposed Christian.

So instead of bearing to starboard on leaving the channel, we swung three kilometres to port to take this old tub in tow. On board were three men in pullovers, crouching under a tarpaulin to shelter from the waves and spray. As we approached, they stood up. Smiling broadly, they caught the thick cable we threw them, made it fast to the bow of their boat, and immediately (wily people, these Greenlanders!) cut the engine to save fuel.

Then we were off again. From boat to boat we talked to one another, shouting at the tops of our voices to make ourselves heard above the engine and the wind; then we started throwing things at one another, laughing wildly. We had a good time, and seals were completely forgotten.

Three quarters of an hour later, some other men hailed us from a vessel moored at the foot of a mountain. My shipmates asked those we were towing to cast off the cable and throw it back to us, which they did without a murmur. They went on, while we made a detour to go and see what the others wanted. They were three more fishermen, not in any trouble in the storm, only they had boiled too many gulls. They passed us the pot, and for half an hour we sat eating and discussing the storm, then set off again, towing them to a point off Fiskenæsset. I could appreciate the Greenlanders' helping one another outside their villages but lamented their inability to get on with the job.

'Look, houses!' Christian shouted.

'Where?'

'In that gap in the mountains.'

'Which gap?'

'There!'

But for a good ten minutes I couldn't see anything.

'Incredible! Are your eyes that bad?'

Finally I could make out a white dot that now and then pierced the fog. The dot grew gradually clearer, and a quarter of an hour later I recognized a house, with others next to it.

'It's Fiskenæsset.'

We arrived at six o'clock in the evening.

'It's too late to go fishing now – let's take a look around,' they suggested.

Everywhere in this village were big wooden frames for drying fish. We passed the church, a low building painted red. After stopping at the store for bread, sugar, milk and jam, we went to visit a family. Empty beer bottles were heaped behind their front door. The family was having dinner: the mother and two daughters sat in the kitchen/living room, while in the next room three men were tucking into a big dish of boiled salmon. They sat us down with them, and we shared their meal, all digging fingers into the same dish. It was the first time I had seen women eat separately in this country. It surprised me, but this segregation of the sexes, and the way we were stuffing ourselves with food and swapping stories, reminded me so much of Africa that I didn't even ask about the custom.

After paying coffee calls on three more houses we returned to the boat, where it was our turn to receive visitors. The little cabin was crammed with people sitting on the floor. Open, laughing faces with plump, yellow, oily cheeks gleamed through a cloud of cigarette smoke. We guzzled coffee, laughed a lot, and played draughts. The gangway was jammed with children. Some of them climbed on deck and jostled outside the cabin trying to get in. Abel tried to get rid of them, but they didn't budge an inch, and all craned their necks to get a good look at me. Every movement I made unleashed a gale of laughter.

There was no electricity at Fiskenæsset, a village of three hundred inhabitants. All the houses were lit by oil lamps; in the streets they used flashlights.

We went to the local dance, which was packed. By the light of the oil lamp, so many young women with shining cheeks gave me obvious come-ons that, for fear of annoying their husbands, I stopped meeting their soulful glances. Only lemonade was on sale at the counter, so Christian took us around to the back of the building to lace our drinks secretly with the *akvavit* he carried in the inside pocket of his coat. A girl happened by and took a swig from everybody's bottle. We went back into the hall but left not long afterwards to rejoin our ship. Outside there was total darkness, pierced here and there by headlights.

The next morning, Monday (the fourth day after leaving Nuuk), we got up at seven. After a bowl of coffee, we cast off from the landing-stage and stopped in a nearby bay, not to hunt this time but at long last to make preparations for the oddest kind of fishing I've ever seen.

I've already said that on board there were tubs full of ropes: fifteen of them. The ropes were blue, and each tub held more than five hundred metres of them, carefully coiled, and equipped at intervals of thirty centimetres or so with secondary lines of thinner rope, each about twenty centimetres long and ending in a hook. What we had to do was join the main ropes end to end out of the fifteen tubs, then pay out this long ground-line, with its hundreds of hooks, under the water.

To start with, we heaved big packets of frozen Norwegian mackerel out of the hold to be cut up as bait. 'Not too big, not too small, but *this* size – yes, that's it!' said Christian. Then, in the drizzling rain, we crouched in front of the first tub and dumped its contents on the deck. We started at one end of the rope, baiting the hooks one by one while coiling the rope back in the tub, but with the two ends hanging out. I discovered why later on. Most of the hooks still carried old bits of rotten mackerel bait. We took these off and stuck on fresh bait, throwing the rotten bits into the sea. This attracted birds, which came hovering overhead, but no one took any notice of them, the work was too absorbing.

When the first tub was finished, we started on the second, again dumping its contents on the deck. But before starting to fix the bait, we secured one end of this rope to one end of the rope in the first tub. By midday all the hooks were baited and the fifteen tubs lined up on the deck, linked by the knotted ground-line.

After this first stage, I wondered how we were going to set about laying this long line under the water. I could imagine the line thrown overboard, to be carried away by the current and lost forever.

We moved another two kilometres away from the village. When we had reached a spot that Christian judged right for putting out the line, we took a thick rope without hooks on it. We tied a yellow balloon to one end, a heavy stone to the middle, and fixed the other end to the ground-line in the first tub, then threw it all into the sea. With the balloon floating and the heavy stone resting on the bottom, we could now pay out our line under water. It didn't show on the surface, because it was held down by the stone; nor was it lost, since its position was marked by the yellow balloon. We spent an hour edging the ship along, paying out the kilometres of baited line. The other end of the line was also attached to a thick, unbaited rope, with another heavy stone and another balloon.

'*Taama!*' (Finished!) my shipmates cried when this second balloon was bobbing on the water.

'The line must stay under water until ten tonight,' Christian informed me.

So stage two was underway. It was one o'clock in the afternoon, and our minds turned to food. We had the rifles, but the birds which had been following us earlier had now dispersed, and the village had long since disappeared among the mountains behind us. We drew alongside a rock coated with lichens and let the engine idle while we went ashore and up into the hills where, for the next hour, we picked and ate *paarnat* before returning to the village.

Back in the boat, Christian at once lit the little oil stove to make coffee. Soon afterward he put a pan on to heat, and the smell of boiling meat invaded the little cabin.

'What's cooking?'

'*Tuttu.*'

'Reindeer again?'

'*Suuuu.*'

This boiled unseasoned meat had already made me throw up twice at Færingehavn during the storm. But we still had some packets of mackerel left.

'We have some fish, haven't we?'

'We do? Where?'

'In the hold.'

When they realized I was talking about the mackerel, my three companions stared at one another in astonishment.

'*Mamanngilaq!*' (It doesn't taste good!) they told me. 'It's only good for bait. But if you'd like some . . .'

So, after boiling the reindeer meat that my shipmates ate unsalted, we put some mackerel on the stove. I was the only one to eat it, again without seasoning. The rest was dumped overboard and the pan thoroughly washed – for the first time – with sea water.

We still had three hours before we went back to our line, and we spent them paying visits, drinking coffee in one house and beer in another. A young man called Pavia offered to come fishing with us. Christian accepted because we could do with a fifth hand for the work we had in store.

It was six o'clock when we returned to get things ready on the boat. As the operation might last well into the night, I wore a second pullover over my shirt, and Abel lent me a third one, very thick, which I put on over the rest. As for my four shipmates, they put oilskins over their trousers and pullovers, less to keep out the cold than to protect the wool from the salt water. We each took a wooden stave with a big bent nail at the end, and a hammer to finish off the fish. There was also a knife handy to cut off their heads and gut them on the spot. Christian assigned the various jobs, then took over the steering, and we were off for the third and final stage of the fishing trip – the longest and most impressive.

We reached the first balloon at six-thirty. Already the sun was sinking on the horizon, its golden rays piercing vermilion clouds spotted with black. It was cold.

Using my stave with the bent nail, I hooked the end of the thick rope tied to the line and hauled the balloon onto the deck. We took it off and stowed it in a corner, then I threaded the rope through a pulley fixed directly above the side of the ship and gradually pulled it in as the boat chugged slowly ahead. The first big stone emerged from the water; we removed that, too, and I went on hauling in the rope, to which clung a kind of seaweed with broad, reddish leaves, dotted with perforations. Standing alongside me were Abel, carrying his own nailed stave; the young man from the village, armed with the hammer and knife and bracing his legs for the catch; and finally Jørgen at the other end, ready to coil the line into the first tub. Behind us on the deck stood empty boxes ready for the fish.

But just what were these carnivorous fish which we were all set to tackle with our hammer, bent nail and kitchen knife?

'They're *qeeraq*!'

That still meant nothing to me, but I was just as keyed up as the others. They all gave me stern warnings to watch out for these *qeeraq* (or sea wolf), which were 'very dangerous'.

When I finally got the line in my hands, we detached the thick rope. My task now consisted of gradually hoisting up the line through the pulley, while the boat moved slowly ahead. The hooked fish glimmered under water; already the first were rising up the side of the ship, dangling from the hooks. Most of them were grey, spotted like a leopard, and big. When they were against the hull and I started to wind them slowly up, my shipmates shouted a warning.

'*Qeeraq!* Mind your hand!'

Abel came forward and stuck the point of his bent nail into

the first one, which kept on struggling. It had swallowed the hook, but Abel pulled it deftly out again, lifted the fish on the end of the nail, and dropped it on the deck. Pavia then gave it a terrific clout on the head with his hammer, cut its head off, gutted the fish with incredible dexterity, and threw it into one of the boxes behind us.

While Abel gaffed the fish one by one and Pavia killed and gutted them, I had a moment to examine one of the severed heads. Never have I seen a fish with such powerful teeth, and set in so many rows. Talk about jaws! If someone had described such a fish before I set eyes on it, I'd have thought he was pulling my leg.

Pavia single-handed was no longer able to gut all the dead fish lying on the deck, so Abel got a second knife and handed me the gaff. Now it was up to me to get the *qeeraq* on deck, while continuing to haul in the line. At first this was no easy job, because as soon as they felt the nail in their flesh, the fish wriggled and twisted violently. The nail didn't pierce their skins easily, so like Abel I had to strike really hard, either in the neck so that the point could sink in deep, or into the ribs or spine to catch on a bone, so the fish didn't fall back into the water as it was lifted.

Within ten minutes, however, I was managing to use the gaff with the same speed and almost the same impassive competence as Abel. Most of the *qeeraq* lay still on the deck once they had caught a violent blow on the head, but sometimes you had to slam the biggest ones several times across the neck. Invariably, they gave death-rattles when the knife sliced into their bellies, and some then released what looked like a stream of urine. Thrown gutted and headless into the boxes, they still keep feebly twitching. The whole of the fo'c'sle was drenched in blood. At seven-thirty the horizon, barred with black and grey clouds silhouetted against a crimson sky, was stained with red, dark as a lake of congealing blood, while the deck of our ship was slippery with the blood of the *qeeraq*.

We threw the heads and guts into the sea, together with any fish we didn't want, particularly some thin ones and others which were big and greyish. A few hooks still held bait, but Jørgen coiled the line into the tubs without bothering about them. Sometimes the line got tangled as it came out of the water; we had to stop to disentangle it, then set to work again. The sea birds hovered and dived over the steaming entrails. A few, seated on the water, were content to follow in our wake, gorging greedily.

The sea was inky black. The dying light of evening still gleamed on the sombre flanks of the heavy swell. At eight o'clock we lit one of the two lamps on the masts. A quarter of an hour later, the second was lit. Stars were appearing in the sky. Towards nine o'clock, from behind a chain of mountains, the moon rose, big, yellow, heavy, almost on a level with the waves. The weather turned considerably colder, and it grew harder and harder to haul in the line bare-handed, but mitts would have got in the way. We all continued to work without relief: I went on hauling up the fish, two others clubbed and gutted them, the fourth kept on coiling the line, and the last steered the boat at a walking pace, only stopping from time to time to allow us to disentangle the line when it was fouled. We were a real production line, but I must admit that we'd have kept moving anyway just to keep out the cold.

It wasn't until ten o'clock that we reached the second heavy stone and the second balloon, which marked the end of a really profitable fishing trip. We returned to the village that night with about fifty crates of *qeeraq* in the hold. After tipping crushed ice over the catch, we ate some boiled reindeer flesh and went to bed.

The next morning we sold the sea wolves at the harbour

depot, where they would be deep-frozen and exported to the United States. I should add that the fish is delicious, even without seasoning.

Christian and the others decided to stay a week longer in Fiskenæsset before moving on to Søndre Strømfjord to hunt reindeer. As for me, I had to return to the capital; in five days' time I was due to board the *Kununnguaq* for Sukkertoppen, one of the most northerly communities in the southern region. So, since I couldn't stay on with the others to share the carefree life of the south, I returned to Nuuk the next day, Tuesday, September 7, with Dr Hels, on the medical corps boat.

The doctor had come to collect a patient – a baby only a few months old, with thick black hair in a crew-cut, bright black eyes and tiny purplish hands. They took him off to Nuuk without his parents. A Danish nurse, tall and skinny, brought him on board; she spoke French. The doctor went to Fiskenæsset only once a month. Emergencies sometimes brought him ahead of schedule, he told me, but unfortunately in such cases the journey was sometimes in vain, because the boat took twelve hours to cover the 148 kilometres from Nuuk.

Dr Hels invited me to dine with him at Nuuk, and when we were about twenty minutes out, he called his wife by radio-telephone to warn her that he was bringing a guest for dinner. 'My wife speaks French better than I do,' he informed me after his conversation.

When we reached their house a little later, their small daughter took one look at me, turned bright red, and started to scream, while her elder brother, aged three and a half, plied his parents with excited questions. The doctor came back into the living room roaring with laughter and repeated one of the questions the boy had asked his mother: Was my penis the same colour as my skin?

Talking French with my hosts while we sat at a well-laid

table, I felt for a moment as if I were anywhere but in Greenland; yet the succulent steak we were eating was reindeer meat. Yes, that's right, this time the reindeer meat tasted fine. In fact, when it's properly prepared and served, and not just boiled, reindeer meat is one of the Arctic's tastiest dishes.

4

The 'Polar Hysteria' of the Arctic Autumn

Thomas Petrussen had warned me at Qaqortoq that, during the autumn, a type of nervous depression called 'polar hysteria' develops in certain individuals with the approach of winter. The affected person alternates from a passive, listless state to an unbridled fury in which he smashes everything within reach, then relapses into a drained stupor. I was told that medicine has no remedy for this weird nervous or psychological affliction, which is triggered by a number of factors. These include the approach of the long polar night, which will last for more or less six months, according to the latitude; the oppressively dreary autumn light; the debilitating inactivity (during this transitional period the onset of the freeze prevents kayaks from staying long at sea, yet the young ice is not thick and hard enough for sledges); the paralysing boredom that results; some obsessional idea (in autumn, people don't know what to do with themselves or with those around them); and the cumulative effect of lack of sleep during the Arctic summer. The Greenlanders I lived with during the summer were very light sleepers – and loud dreamers; sometimes they would even walk as well as talk. Because they don't have thick enough curtains to keep their rooms properly dark, their sleep throughout the summer is troubled. Some people, whose nerves have been badly jangled by this kind of near-insomnia, can hardly bear the autumn and the slow approach of the long months of darkness, and are prone to fits of hysteria.

It's not surprising. During the last war there were French soldiers living four metres underground in the fortifications of the Maginot Line who suffered attacks of hysteria brought on by 'darkness and inactivity'. These attacks were dubbed 'concretitis'. Generals, being always busy, were not affected.

In the Arctic these hysterical attacks affect natives and foreigners alike.

I arrived at Sukkertoppen (Maniitsoq, 182 kilometres from Nuuk) on September 12 at eight o'clock in the evening, in pitch darkness, and was taken in by Erik Rasmussen, a stolid, taciturn Greenlander, about thirty-five years old, and a father of seven. He was on vacation from his job as a telegraph operator and spent all day indoors, bored stiff. His wife, who looked after the kids, worked at the local fish factory. The family lived in a three-roomed flat. In the kitchen, two upholstered benches along the walls allowed the whole family to sit at the table. Erik and his wife slept on a mattress on the floor in the room by the hall, with one or two of the smallest children. The others slept in the other room, crammed into double bunk beds, two or three to each.

The flat was in the right wing of a long building which housed four or five other Greenland families, with a Danish dentist and his wife on the far left. The apartments all had the same number of rooms, but the Danes, with only two children, were not as pressed for living space as Erik. In his place it was impossible to sleep well, because of the smell of children's urine and excrement that lingered even in well-laundered sheets, and the continual din. That alone would shatter the steadiest nerves. In the middle of the night the smallest of the brood, aged one, tottered and crawled from one room to another, imitating the sound of a motor car. Then he cried and fell silent by turns. Nobody bothered to put him back to bed. Lost in the dark, he howled louder and louder, often until morning.

At breakfast on Monday, September 20, I was astonished to see Erik lose his temper, not with the little one who had kept us awake, but with a daughter of three who hadn't done anything. He didn't reply to an idle question I asked him in an attempt to start a conversation and calm him down. Yet he did seem quite calm – it was just that he no longer talked to anyone. His wife, perhaps guessing what was in his mind, gave me a wink. I still didn't understand. Erik had not had a drink.

He went back to bed after breakfast but didn't sleep. At lunch-time, his wife called him to the table. He got up and sat down, wild-eyed but still not speaking. Everyone knew now that there was something the matter. The children avoided him and hushed up. We were supposed to go fishing together, but given Erik's mood, it was out of the question. There was no change after the meal.

That evening about seven his wife emerged from the kitchen and asked me:

'Have you seen Erik?'

I hadn't seen him all afternoon; his wife panicked. She ran outside, then back in, then out and in again, very upset. Little Alfred, aged two, answered a question from his mother with one word: *politi*. Erik, who had five daughters and two sons (the latter two were the youngest), was more affectionate with the boys and often took Alfred out. It appeared that they were both out walking when the police arrested the father and left the child wandering loose in the street. At the word *politi*, Erik's wife rushed out for the third time but soon came back, red-faced. There was a striking contrast between the vivid red of her face and the sallow skin of her youngest child, whom she carried in her arms. We sat down to dinner, but she ate hardly anything and paid no attention to anyone, not even the children. She kept muttering: 'Maybe he'll have to spend the night there. It's the first time – yes, the first time *this* has happened to him!' Then she rushed off again like a madwoman. I looked after the children, but I was ready to go to the police station if the wife returned without her husband.

Just then Erik came home, brought back by two officers in a police car. Seeing him covered with blood, with his face all swollen and skewed to the right, I asked the policemen what had happened.

'He went into a house and started a fight with a man and his two sons who had done nothing to him. He left there and started insulting people in the street, and boasting that he was the toughest man in town. Keep an eye on him, don't let him go out tonight. Let him sleep it off. He'll get over it.'

And the policemen left, dispersing the crowd gathered outside.

Without looking at anyone, Erik went straight to the toilet and then into the kitchen, where he began to eat with his fingers, like a starving man. The children kept quiet and stared at him fearfully. He opened and drank one of the bottles of beer we had bought the day before for our fishing trip. His wife served him another great pile of food, more than the nine of us had eaten together (boiled fish, without vegetables), then sat beside him, tears streaming from her eyes, overflowing with tender concern, her hands raised towards her husband's chest as if ready to protect herself if he made a violent move, or to embrace him if he turned gentle, as he did. Only then did Erik speak:

'There were three of them! *Pingasut!* Three! And me, alone, *ataaseq!* Alone!'

Then, cramful of food, he got up and announced:

'Now I'm going to fight!'

We did our best to dissuade him, but it was no good. So, in order not to cross him, his wife and I went out with him in the rain.

Fortunately, he didn't start running. He even stopped once to buy some cigarettes, then we went on walking. But because of the rain and the children left at home, his wife decided to go back, after asking me to stick with Erik. So there I was, walking through the night beside this man who seemed to have lost his mind, and who kept repeating as he walked: 'I'll kill him! I'll kill him!'

In the little valley by the cemetery, we crossed a bridge on our right, then Erik entered a house. I followed him, and quickly surveyed the occupants – a girl sitting at the back, an old woman, two youths, and an older man wearing a leather jacket daubed with blood. So this was where Erik had come to start a fight. The people stared at us, flabbergasted, saying nothing. Erik shouted threats at them, then turned on his heel. 'Next time I see you,' he kept growling on our way back, 'I'll kill the bunch of you!' As he reeled from side to side, people turned round to watch. He pissed against a wall, staggered round the house, then circled three or four times round a tractor parked across the street. Finally we went back inside, and he stretched out in the armchair where he usually sprawled after dinner.

The next day Erik didn't get up until past noon, and then it was only to make room for his wife, who was sweeping. He went off at once to sleep in the bedroom and didn't get up until ten that night, having had nothing to eat all day. After sleeping round the clock like that, he remembered nothing of what had happened. The next day, Wednesday, he was his old self again, taciturn as usual, but more cheerful.

The Arctic autumn had no ill effects upon me personally. My role of close observer, which kept me continually alert, together with my constant longing to get further north, and knowing I was only staying a short while in Sukkertoppen, may account for this. Unlike the inhabitants, I wasn't marooned in that narrow little world. Yet the weather was sometimes so dreary that my morale sank pretty low. Some nights I woke up in an agonized sweat, overcome by a terror I couldn't explain. It was in autumn that I had the most nightmares.

The first snow fell on September 15. So thick were the flakes, you'd have said that all the white birds in the world were shedding their feathers. Between the slanting raindrops, sodden snowflakes plummeted straight down and melted as soon as they touched the ground. The rain poured down again two hours later, heavier than ever, and the snow squalls grew more violent.

On the fourth day after that first snowfall, the pools of standing water trapped in fissures on the plateaux began to freeze over. Their surfaces looked like plate glass, the edges like solidified candle wax. Five days later, a small lake at the foot of the mountain also froze over. The sea was still clear of ice, but the darkness lengthened from day to day.

On the night of the day the first snow fell I was frightened by a bizarre phenomenon. I was walking home alone and the night was still. Suddenly looking up, I saw long white streaks whirling in the wind above my head. It was like the radiance of some invisible hearth, from which dazzling light rays shot out, streamed into space, and spread to form a great deep-folded phosphorescent curtain which moved and shimmered, turning rapidly from white to yellow, from pink to red. The curtain suddenly rose, then fell again further on. The wind shook it gently like an immense transparent drapery carried by the breeze and drifting on thin air. Its movements were now regular as an ocean swell, now hurried, jerky, leaping and tumbling like a kite. There were continual changes in the intensity, the motion, the iridescent play of colours and the ripplings of this strange, gigantic veil that floated through the night sky. I stood watching it for ten minutes, stunned and fascinated. Not knowing what it could be, I rushed home and babbled something about it to my hosts, who didn't bother to go outside but informed me that I had just been watching the aurora borealis. I went out to watch it again for a long, long time, so much more impressive was it than anything I had ever read about these polar auroras.

The northern lights shine out in a black sky and are just as likely to move eastward as westward. They are as bright as the full moon. Sometimes they look much the same as ordinary clouds, and are distinguished from them only by their rippling movement and phosphorescent radiance.

My three-month visa had just expired and the chief of police, a Dane, asked to see me in his office. He had received a telegram about me from Denmark and, explaining that my papers were not in order, asked me what I intended to do about it.

With Alfred, son of my host
Erik Rasmussen, in Maniitsoq

'Renew my visa.'

In Greenland the police never act on their own initiative; the chief had to write to Copenhagen for a decision. How long would that take? You never could tell . . . As I only had one week left in Sukkertoppen before the arrival of the coastal vessel which would take me further north before the ice set in, the commissioner advised me to apply to his colleague in Holsteinsborg as soon as I got there.

I had to renew my visa five times in Greenland. Each of these visas extended my stay for three months but did not allow me to work. My adoptive father in Paris was getting worried, wondering how I could manage in such circumstances. He gave me help I will never forget by sending me six hundred kroner a month, which I used mainly for the fares on the coastal vessel, since most of my hosts, including Erik, asked nothing for my board and lodging. I contributed to the food by going fishing or by buying provisions for us all from time to time.

That's how I once came to buy porpoise for our dinner. It was a big slab off the back, sold with thick lumps of fat. The flesh was as black as coagulated blood. When he saw it, Erik exclaimed, 'Tomorrow we'll have a full meal, with dessert!'

'What dessert?'

'Why, *paarnat!*'

And the next day, a Sunday, we set off early in the morning – Erik, his wife, the children and me – to gather *paarnat*. We didn't get back until three in the afternoon, with a bucket and plastic bag full of berries.

That evening, a thick slab of boiled porpoise was set on the table, flanked by huge lumps of steaming fat curled up at the ends, and accompanied by a dark broth made with a base of blood and containing rice, carrots, and potatoes. Then came the first dessert: *paarnat* berries in a great oval dish. The second dessert, served immediately afterwards, was a mixture of *paarnat*

A view of Maniitsoq

and sugar over which we poured condensed milk. Then, stomachs full, we went outside for a breath of air while the children stayed in the house and played at *timmisartoq* (helicopter) with the porpoise's three-pronged vertebrae. Greenland children play with all animal remains: after eating poultry, for instance, they play with the breastbone, flipping it across the table by pressing with a finger on the lowest part, the sternum.

I left Sukkertoppen on September 26. That town marked the end of my stay in southern Greenland – a region with no huskies and no sledges, a land of grey landscapes without ice. I had still to discover the cold white wilderness with its Eskimo seal hunters which I had never ceased to dream about: the Far North.

PART III

Hibernation

I

Sisimiut, the Gateway to the North

Since six o'clock we had been in the Davis Strait skirting a series of snow-covered peaks almost blue at the base. In the afternoon we turned off into the tranquil waters of Søndre Strømfjord (Kangerlussuaq) to let off at the far end of this fjord passengers bound for Denmark, and to pick up those who had just flown in and were waiting for the boat to take them to their various communities. Søndre Strømfjord, an American base, has a hotel with fifty-six beds, as well as a small civilian airport, the most important in the country, where planes flying the North Pole route to Los Angeles touch down.

We set off down the 190 kilometres of the fjord, the longest on the west coast, its narrow, twisting channel bordered by imposing chains of mountains whose jagged peaks were reflected in the blue and green depths. The sides were choked with winding glaciers, most of them stationary and blackened by moraines. As we entered the imposing and romantic spectacle of the fjord, the film being shown on board was interrupted by solemn recorded music. And while this music was playing, amplified by the steep walls of the fjord, our handful of passengers was presented with navy certificates testifying that we had travelled inside the Arctic Circle, which we had just crossed. This distribution of printed forms struck me as so grotesque that I didn't bother to collect mine, preferring to savour the strange thrill of that striking landscape.

We sailed on for hours, not reaching the other end of the fjord until nighttime, when the vessel moored about one kilometre from the shore. Lights shone in a gap to the right, but the airport was still twelve kilometres away. There was a boat to ferry passengers to and from the airfield up an arm of the fjord too narrow for our own ship, but it operated only during daylight hours, so we all spent the night on board and had to wait till morning for these passengers to disembark.

At first light the ferry picked up those who were going to catch the plane, then returned two hours later with the new arrivals. A fishing boat came alongside our vessel with reindeer carcasses and new skins stretched on the yard, and the ship took on some meat for the passengers' meals. We left Søndre Strømfjord at nine-thirty, in rough weather. The sea was green and foaming, and a cold wind carried the first drops of rain. As we moved out into the high seas, the ship caught the full force of the storm and pitched and rolled violently until our arrival at Holsteinsborg at nine o'clock that night. A barge came out several kilometres from the coast to pick up passengers for this village. Barge and ship were lurching so violently, hove to on an angry sea, that the women had to be helped along the gangway. One of them slipped, and she and her luggage would have gone to the bottom had she not been caught just in time. Soon night fell on the little shallow-draught boat. We clung on for dear life as it battled towards the village, while behind us the *Kununnguaq* was no more than a great luminous halo on the sea.

The barge deposited us on a quayside packed with people, baggage and goods which would now be ferried to the coastal vessel. Two taxis were waiting, so I took one of them to deliver a letter to Lars Peter Olesen, my friend Erik's brother-in-law. Lars Peter was a teacher at the Knud Rasmussens Højskole, the local high school, and lived on the premises at the other end of town.

The entire town of Holsteinsborg (population 1,750) is strung out along one main street, down which we promptly set off. The place was deep in snow, and as the taxi's headlights shone on sledges standing outside doors, I caught sight of eyes blazing in the night – my first fleeting glimpse of Eskimo dogs.

As soon as I set foot in the schoolyard, some girls shouted, '*Kinaana*?' (Who's that?) The whole yard was in an uproar. One of the girls went to fetch Lars Peter, who was also very surprised to see me, but like any good Greenlander politely concealed his astonishment. In spite of the letter of introduction from his brother-in-law, he made it clear – trying not to speak Eskimo – that he couldn't ask me in or put me up. Whatever his reasons, I simply record that this was the first time I'd been refused hospitality in this country, but the refusal came from a Greenlander who lived more or less in European style and was trampling – or so I am tempted to think – on one of his country's most sacred traditions.

I got back into the taxi, intending to return to the *Kununnguaq* (which would be at anchor until the next day) and go straight to the next village, but at the intersection I changed my mind and asked the driver to drop me there. Holsteinsborg is a great fishing centre, despite its Eskimo name, Sisimiut (The People Leaving by the Foxes' Holes). It is also the beginning of the 'dog line', the junction of two ways of life, the gateway to the Far North, a border town whose mentality I felt I ought to get to know, as a step towards understanding those who lived further north. After all, Lars Peter was not the sole inhabitant. So I got out of the taxi and headed for the nearest house. I knocked at the door, it opened, and a man and a woman, both young and wearing grey pullovers, appeared.

'I've just arrived, and I'm thinking of staying a while in Sisimiut. Can you put me up?'

'Of course. Come in.'

I paid the taxi driver seventeen kroner and brought my luggage into the house.

The couple sat me down beside their four children: a boy, Aqqaluk, of nine, and three girls – Risa, Rigmor and Nuka – who were eleven, six and four, respectively. The smallest told me she had made a trip on the *Kununnguaq* not long before with her mother and had seen me then.

'You boarded at Nuuk and got off at Maniitsoq. I wanted to speak to you, but you were walking too fast for me.'

'When we got back,' her mother added, 'she talked so much about you, I had to buy her a black doll.'

'I'm glad you've come to live with us,' the little girl continued, and came up close to run her little fingers through my hair.

We sat drinking coffee until eleven o'clock.

'Tonight you'll be sleeping in the living room,' the father told me, 'but from tomorrow you can have one of the two bedrooms upstairs.'

Then, glancing at the sofa where I was to spend the night, he observed to his wife that my feet would stick out because I was so tall.

'Stand up a moment,' he asked me.

I stood up, and he stretched his arms to measure me – first from my head to my knees, then from my knees to my feet. Crouching, he then carefully measured the same length along the improvised bed.

'It'll be all right if you fold your legs a bit,' he concluded, getting up amid the laughter set off by these words.

My friendly hosts were Ludvig and Jakobina Andreassen.

I had an atrocious night because of the fierce barking of dogs under my window. It sounded as if a score of demons were tearing themselves to pieces, so great was the commotion. I got up to look out of the window, and found myself watching a raging battle. It was an indescribable sight. Two of the dogs had been mating, which must have started the trouble. They remained joined at the hind quarters, with some fifteen others milling around them, barking furiously and relentlessly savaging the unfortunate couple, who, trapped, had to keep turning in a circle to defend themselves. There was a horrifying free-for-all; you heard nothing but hoarse spasmodic pantings, savage yelps, murderous snarls and the snapping of jaws.

Drawn by the din, other dogs came running from all over the village. Not knowing whom to attack, they bit at random into the pack. Sinking their teeth in, the beasts piled up on top of one another to form a living heap of flesh and fur that writhed and then collapsed. Hot, steaming bodies flew in all directions. Most of them fought to kill: they went for the throat and hung on grimly, their clamped jaws grinding where they gripped. The unlucky couple fought back bravely, but the contest was too one-sided. First the male and his female reared up on their hind legs, fur bristling, teeth bared; then they were thrown to the ground but fought on from that helpless position, still trying in vain to disentangle themselves. Savagely harried by their adversaries, no longer knowing where or what to strike, they bit at each other, too, without pity. Still welded together, they managed to get their backs against the wall of the house, a better position to confront their assailants, now divided into two packs.

After ten minutes of hellish uproar under my window – a pandemonium of flying claws and raging fangs – the attackers suddenly broke off, crouched down, gazed at their whimpering victims, and began to utter plaintive howls, their muzzles pointing to the sky. In the distance, the voices of all the other dogs in the village joined in chorus with the pack outside. Their cries soon became one long ululation, like sirens wailing; then, little by little, they subsided, only to begin again.

Needless to say, these animals are not domestic pets and do not sleep indoors. They spend the night outside in the snow and bitter cold, even when the temperature falls to more than forty degrees below zero. Many of them weigh a good fifty kilos and can eat as much as a man. An average-sized husky demolishes two to three kilos of food a day, if not more. A Greenland hunter today generally owns fifteen to twenty such dogs. You need from eight to thirteen to make a good team, though at the beginning of this century it took half that number.

The dogs at Holsteinsborg, when I took a look at them on the morning after the savage fight which had kept me awake, bore no resemblance to their descriptions in books, which portray them as strong and healthy. On the contrary, they were emaciated, ill-treated, and largely neglected by their owners. These scraggy curs, apparently ownerless, roamed all day long round the fishing port, waiting for someone to throw them some scraps of food; they were more likely to get a boot in the ribs. Occasionally someone would toss them an unexpected lump of rotten fish, which they would pounce on ferociously, fighting among themselves. Across the street, four slaughtered huskies were hung up next to a fishing net, their necks stretched by the ropes, jaws drooping slightly, gaping and bloody.

In this community where fishing has become the principal occupation, there was obviously less and less need of husky teams. Those people who enjoyed good fishing in summer and autumn (the cod-fishing season here runs from May 3 to October 29) had enough money to live through the winter by buying imported food at the local stores. They no longer needed to hunt for food with a team of huskies as they used to do.

Yet, not everyone in Sisimiut was a fisherman. There were still a few hunters there – those who had no boats of their own. These men did not abandon their *qimmit*: they fed them well, because the dogs were still of some use. As for the hanging dogs, their skins would be used to make trousers.

'All these beasts,' Ludvig told me, 'even those that still belong to hunters, are lean and scrawny at the beginning of winter.

During the summer, they do get fed, but *immannguaq* (just barely enough). Some people don't bother to feed them at all. When there is no *aput* (snow) they aren't needed. Some hunters take their packs to uninhabited islands for the six months of summer and bring them food only every two or three days. Come September – now, in fact – they're brought back here, all skin and bone. Then they're fed regularly, to give them enough strength for their hard winter's work. You've not yet seen all the *qimmit* in this town. Most of them are still out on the islands.'

These dogs take care of the town's cleaning and sanitation. As soon as a woman comes out of her house to empty her privy bucket, the dogs follow her and fling themselves upon the contents. They also eat each other. If during the previous night's battle one of the beasts had been seriously wounded, the rest would have made a meal of him. Puppies with careless mothers are polished off in a couple of mouthfuls, which may explain why bitches about to whelp are sometimes allowed to sleep indoors in the entrance hall, where they live for a few weeks with their litter.

Most blood-curdling of all, these dogs also eat human beings! They sometimes go for small children, if they go near a pack with no whip or stick in their hands to scare the dogs off. These beasts even attack grown men: one of them barks and pounces with bared fangs, then immediately dozens more will come howling for blood. You're trapped without club or whip, in the thick of a maddened pack of two or three hundred starving dogs grown deaf to the voice of man. It's not unheard of in the north of Greenland to meet a grieving mother who will tell you, recalling her most painful memories: 'I had nine children altogether, but two of them were eaten by our huskies.' For they snap up babies whose parents leave them outside, or with the doors not properly closed. Possibly that explains why, all over the North, the entrances to houses have both inner and outer doors. One could counter that in the villages of the south where there are no huskies the houses still have their two entrance doors; but in those areas it was not until the eighteenth century that – because of their voracious appetite for the lambs and chickens the Danes were introducing into the southern regions – the dogs there were exterminated.* If the use of double doors has outlived the dogs, it's simply because centuries of custom don't disappear overnight.

When a human being falls prey to a pack of dogs, it is the Greenland custom to bury an empty coffin. The Danish police slaughter the dogs which joined in the feast, and their bodies are thrown into the sea to prevent the inhabitants from eating them in their turn. All the dogs in the village are rounded up and examined to detect the culprits. Curiously enough, when that happens, every dog owner – Dane or Greenlander – springs to his dog's defence. However, those dogs with a gorged look or blood-specked muzzles are suspect, and they are shot in the presence of their innocent companions. It hardly seems likely that the innocent are really aware of the dead dogs' crime and the reason for shooting them. In any case, these measures do not prevent recurrences. But I don't want to get ahead of my story and talk about things I hadn't yet seen for myself. We shall see these dogs again on many occasions – especially in the villages where they live in the hundreds, and sometimes outnumber the inhabitants three or four to one – and gain a better understanding of their unusual behaviour.

In the afternoon I went to see some of the islands where the huskies were abandoned for the summer. Markus Byarnassen, the Danish principal of the primary school, took me in his motorboat. As we approached one of the islands, the dogs

---

* Kaj Birket-Smith, 'Ethnography of Egedesminde District, 1918', *Meddelelser om Grønland*, vol. 66 (1924).

began to bark and to rush down the rocky slopes, thinking we were bringing them food. Others on the next island watched us chug by without moving. Not far from them lay the uneaten bony head and backbone of a blue shark, which had been brought to them a few days before. These dogs have to be especially tough to survive the treatment inflicted on them during the six summer months, and remarkably inclined to go on loving man – if love him they do.

Apart from the huskies, the sledges, the snow and the long polar night now beginning, this community was in no way different from those in the south, with much the same daily life. Here, too, drink played a major part, and the 'rural exodus' had created the same havoc. One of the villages I visited with a party of Danish teachers was Uummannaarsuk, in the parish of Holsteinsborg. There we found not a living soul, not one dog. All the inhabitants had been hunters. They had abandoned their village en masse to become fishermen in Holsteinsborg, where most had instead become unemployed. A Danish schoolmistress in our party, who had adopted two children of the Uummannaarsuk 'refugees' to save them from starving to death, picked up an old seal-oil lamp carved from soapstone, chipped at the rim, in one of the abandoned houses. Behind the doors were casks for storing drinking water or ice to be melted down. In some houses we saw platforms used as beds; in others, where there were no platforms, the people slept on skins on the ground. In a loft we discovered a load of reindeer skins. The doors of the deserted dwellings banged in the wind, as if in the wake of some murderous raid. It was a sorry sight. I have never seen anything like it in my life. Lying about in the village we found scraps of reindeer skin, and an old pair of women's breeches that another Danish woman picked up. And so, after the ravages of centralization, the abandoned villages bring out a taste for – looting.

Knud Oleson, my host's father-in-law, came to pay us a visit. Fifty-five years old and born in Thule, where his father was a Sunday school teacher, Knud had a rather Mediterranean look. There was something attractive about his curly hair, greying at the temples, his little moustache and rounded cheeks. He greeted us with a shy smile and, instead of joining us at the table, sat down with Risa, his eleven-year-old granddaughter, and quietly asked her my name, my age, and whether I had any brothers and sisters. A little later his wife arrived, older and smaller than him. Then he got up and they came and sat with us. Was it because of Knud's reserve and the restrained curiosity in his eyes that I sensed in him a man belonging to an ancient race, an aristocracy? His politeness was exquisite, his manners simple and agreeable. His smile was not forced, but calm and gentle. He had been a hunter of seal, narwhal, moose and whale, until a hunting accident put him in hospital for twelve months in Copenhagen, which he regretted not having been able to visit. 'All I saw there was the hospital and the white-coated doctors and nurses,' he said with a smile. When he returned, it was with a slight limp and he was unable to continue hunting, so he had to take work as a warehouseman in the port. Father of nine children – five boys and four girls, including Jakobina – he now had twenty-six grandchildren.

The conversation turned to the preparations for Rigmor's next birthday, when she would be six. Six days before the event, the last details were being arranged. With grandfather and uncles chipping in, Jakobina and her mother conferred: 'Now, let's see, how many kilos of coffee, of sugar? How many packets of biscuits? And what about the evening meal? What wine shall we have, what dessert?'

All the time she was talking, Jakobina was serving us coffee, after which she brought in some reindeer fat. Each of us cut off a piece and put it in his hot coffee with or without sugar, laced with *akvavit*; I did the same. As it melts, the fat forms little oily circles on the surface. After the coffee is drunk, there is a little fat left in the bottom of the cup, which you take out with a spoon and swallow with a lump of sugar.

'How do you like it?' Knud asked me.

'It's very good!'

'Well, then, you must come and have some in *my* house.'

I was to see Knud almost every day after that. Each time we sat down to this delicious mixture of coffee-*akvavit* with reindeer fat, then little glasses of liqueur, while my host cheerfully recalled his hunting days, and his meetings with the great Danish-Greenlander explorer Knud Rasmussen, a friend of his father's. His admiration extended also to the ethnologists working in his native region of Thule, and in particular to Jean Malaurie, who through his many activities as writer, filmmaker, and lecturer has called international attention to the serious problems facing the Arctic minorities under white administrations. It was in fact through Jean Malaurie's initiative and under his chairmanship, with the help of the late René Cassin, Nobel Peace Prize winner, that, for the first time in all their long history, Eskimos from Siberia, Alaska, Canada and Greenland were assembled together. This gathering took place in France in November 1969, and through it the Eskimos' problems received their first international airing. The Greenlanders with whom I have kept in touch have always been grateful for that French initiative.

On Tuesday, October 5, I was awakened by strange voices talking noisily downstairs, and the sound of children crying. It was Rigmor's birthday, a celebration that began at daybreak. From early in the morning, people came in one after another to drink coffee, and left with the Eskimo *qujanaq*, thank you, while others arrived alone or in little groups. The presents they brought for the little girl piled up in a heap on the sofa: dresses, *kamiks* of sealskin lined with dog fur, trousers with straps to pass under the shoes, scarves, chocolates, and so forth. Most people didn't bring anything. I went out to buy some film for myself and a doll for Rigmor. Smiling broadly, the mother greeted her visitors, waited on them, and bustled to and from the kitchen with her three coffee pots, each taking its turn on a table laden with biscuits, cakes and reindeer fat. All day long Jakobina was busy with one thing only: making coffee. Many people came back three, four, even five times during the day. I should think the whole town, or at least three quarters of it, must have paid a call, for this incredible coffee serving, which started early in the morning and never once stopped, didn't end until half-past six in the evening. Jakobina had asked for a day off from work to organize the celebration, but she was so exhausted afterwards that she had to stay at home for another two days.

The evening meal was limited to a few relatives and close friends. Eight chairs stood round the big table where we took our places. A white tablecloth had been spread, with napkins folded into bishops' miters on the plates. To my right, Knud sat at one end of the table, facing his son-in-law Ludvig at the other end. Opposite me, Knud's elderly wife sat smiling happily between her husband and one of their sons, who was married to a schoolmistress from Nuuk. On this son's other side was one of his great friends, a fisherman. The two seats next to me were empty. Jakobina did the serving. Eyes sparkled at the sight of the two five-litre bottles of red and white Spanish wine which she set on the table. We tucked into roast seal cut into succulent slices on a great oval dish, accompanied by red cabbage, carrots, potatoes and gravy. Each guest had four glasses filled with red wine, white wine, beer and fruit juice, to be drunk as they pleased. A fifth glass was reserved for liqueur. We were on our

third helping of roast seal when we were joined by the schoolmistress in a maternity dress, the fisherman's wife and a married
couple, of whom the husband – hair plastered to his forehead
with sweat – started eating at once without waiting for his wife,
who was left standing because there were too few chairs. I
offered her my own, but Ludvig protested and made her sit in
his place. Her low-cut dress revealed her ample breasts, which
rose and fell in time with the flaring of her nostrils. The women
who arrived later installed themselves in armchairs or on the
sofa, and dined off a small, low table. Soon all the other women
left us to gather around this little table, where even Knud's wife
joined them. The men started talking freely among themselves.
Then we were treated to a fruit salad, followed by an odd mixture of *paarnat* berries and cod's liver.

After this double dessert, Ludvig brought out gin, 'Long
John' whisky, and more beer and fruit juice. Everyone helped
himself, Knud beamed broadly, his son fell silent, the women
clucked away, the fisherman talked loudly, banging his fist on
the table, and we were enveloped in a cloud of cigarette smoke.
Someone played the piano; we danced. I left the party at about
eleven-thirty and went upstairs to sleep, my head buzzing. I was
in that unbearably queasy state that makes you swear you'll
never take another drink again. And so, on the pretext of cele-
brating the birthday of a little girl who hadn't been seen at the
table all evening, the adults had organized a treat for themselves. These great blow-outs, if they are to be successful, clean
out the family purse and drain the parents physically, too. Like
Jakobina, Ludvig couldn't work for two days after the banquet.

The next morning was cold, sad, sunless and grey: a real Arctic day. It had been snowing for several days, turning the whole
landscape into a white wilderness. Hail was now falling, and a
strong wind sent gusts of it rattling against the windowpanes.
Nature was a blur: You could no longer make out the shapes of

houses or the great chain of mountains opposite the house. We breakfasted listlessly on the remains of last night's dinner, and then a huge reindeer bone was brought in, which we scraped with a knife to get the last shreds of meat, and we ate the marrow. Overwhelmed by the weather, we drank beer, then coffee, then sherry with cakes. A slow, relentless tide of sadness and apathy came over everyone; Jakobina slumped onto the sofa beside her husband and rested her head on his shoulder.

That night there was nothing left in the house to eat except the guillemots we had shot a few days earlier. We took some of these, plucked them, and put them on to boil. But the birds were only half cooked. There was a pool of blood in the big common platter where we dug into the stringy flesh with our fingers: blood ran down our hands and wrists and smeared the children's faces. The guillemots' hearts, which the children shared, were raw. The people there nearly always eat birds in

The ice fjord of Ilulissat

the same way: everybody takes one and starts with the thighs, then with his fingers pulls the flesh off the ribs, from the shoulders to the parson's nose. The carcass is then torn open for the liver and lungs. Finally, the heads and gizzards are scooped out of the pot.

I sailed again on October 11 at midnight on the *Kununnguaq*, which was making one of her last voyages before the ice set in. I was travelling even further north to Jakobshavn, looking for a village where I could hibernate for the six months of the polar night.

2

Mitti of Ilulissat

The closer to winter it was, the fewer the passengers in the four compartments under the poop. On this voyage I found myself sharing a compartment with only four travelling companions, a couple and their two children. The number of cabin passengers was also very low at that time of year.

At breakfast in the cafeteria next morning, two Jehovah's Witnesses came and sat at my table. Chris, a Dane, and his English wife Joan had been living for a year in Jakobshavn, where they were having a hard time converting the inhabitants. Their aim, they told me, was not so much to establish a congregation as to preach the good news.

'But Greenlanders travel so much and change villages so often, we have to follow the ones we've already contacted from village to village, in order to keep up their Bible studies. We're on our way back from Godthåb: all along the way, both coming and going, we leave the ship at every stop to see old 'students' of ours who were living in Jakobshavn less than a year ago. We'll be seeing others at Egedesminde, the next stop.'

I was sceptical about their chances of making many conversions between now and the end of the world, which they said was close at hand – especially among a people already deeply Christian and entirely Lutheran.

'But again, it's not necessarily a question of conversion,' they explained. 'We simply want the whole world to realize that we are the only true witnesses of the living God. Only those who hearken unto us and follow our example in preaching from house to house shall be saved; all the rest, whatever their religion, will perish with Satan in the battle of Armageddon, which God will fight against him – a battle we are expecting at any moment.'

Such fanaticism made me smile a little, but our differences didn't keep us from becoming friends.

A young Roman Catholic priest, a Greenlander, tall and thin, wearing trousers and an anorak, came in and sat down beside us. He too was going to Jakobshavn, to say mass in the house of the town's one Catholic, a Swiss married to a Greenland woman.

'There are only three Catholic priests in the whole country, and we live at Godthåb, which has a congregation of about a dozen. At the moment, one of us is in the United States taking flying lessons and courses in aircraft maintenance. When he comes back, he'll be in charge of the small plane we're going to buy; trips by boat are getting too expensive.'

Aged about thirty, Father Finn Lynge spoke French like a Frenchman, fluent English and Italian, as well as Danish and his native tongue. Naturally, he also knew Latin and Hebrew. This simple man, whose speech and manner were free and open, without a trace of starchiness, despite his vast learning, was an exceptional person, one of the most impressive I had met there. I was curious to know how he had set about undertaking such extensive studies and become fluent in so many languages – and especially how it was that he had become the only native-born Catholic priest in a country of Lutherans.

'I left here when I was very young to study in Denmark, where I was converted to Roman Catholicism. Then I felt a call to the ministry, and a Danish priest advised me to go to France if I wanted a good religious training. I spent several years in a seminary near Lyon before going on to Rome. Though French seemed hard at the beginning, I soon got the hang of it, thanks to my knowledge of Latin. On the other hand, Eskimo, which I'm now studying seriously, seems to me one of the most difficult languages in the world, because it has such delicate shades of meaning. I speak it, but I can't claim to write it correctly.'

And from his anorak pocket he took a textbook which he was studying in order to learn his native language all over again.

'I've always liked everything French,' he told me.

'The language or the people?'

'Both, because, like them or not, the French are always interesting and always themselves, no matter what other people may think, and I find that admirable.'

When we reached Egedesminde at half-past noon, Finn Lynge (as everyone called him, without the formality of 'Father') wanted to visit the new Protestant church, built in modern style, and I landed with him. We walked for a while along the narrow street that climbed to intersect a road opposite the old stone house of Niels Egede, a local shopkeeper and the son of the first pastor. His big supermarket occupied the whole ground floor of the warehouse, a long black building, the biggest in the town. With its overhanging eaves, the new Protestant church looked more like a farmhouse than a church. It didn't even have a cross: this stood on top of a little bell tower a few metres away from the church itself.

While Finn Lynge went into the church, Chris and Joan, the Jehovah's Witnesses, joined me and took me off for coffee at the house of one of their 'students', a woman of eighty-six, kindly and hospitable. The flock of children who had followed us right up to the door of the house broke a pane in the window as they jostled for a glimpse of me inside. There was no Bible study: it was just a visit.

After leaving the old lady, Joan bought a flat-iron at the store, and we went on board again. A Greenland woman who was boarding the boat stopped and shook my hand, saying:

'I've heard a lot about you. What a pity I didn't meet you just now in town! But next time you're in Egedesminde, be sure to come for coffee at the "Middag Kaffe", the coffee shop I've just opened. Now don't forget – the "Middag Kaffe". '

We left at a quarter past two that afternoon. The cliffs bordering the channel gradually receded and then disappeared, revealing all around us extensive views of a landscape less mountainous than in the south. Now we saw only chains of small islands flanked by long trails of mist. We passed icebergs which were growing in both size and number. The ones in this region come from the frozen fjord of Jakobshavn, which produces more than anywhere else in the North. Towards five o'clock, we saw to our left the island of Disko (Qeqertarsuaq), a great white mass, separated from the rest of the country. As we drew level with this island, I could see on our right the famous Jakobshavn fjord, whose mouth and the sea around it presented a staggering, chaotic view of the frozen chunks broken off from the ice-sheet not far away. Behind these enormous white mountains appeared the community of Jakobshavn, whose Eskimo name is Ilulissat (Mountains of Ice). At half-past six we entered the harbour, which was packed and busy in spite of the darkness.

On the quayside, children helped carry some of my luggage. We followed my new friends Chris and Joan, who had suggested I leave my bags with them while I looked for a place to stay. Their house was one of the first you came to in the village, nestling at the foot of a steep slope. With the snow swept away by the wind, this rocky slope was covered with a layer of ice which had me slipping all the time. As soon as I lifted one foot from the ground, the other would skid off all on its own. On the way down, I threw my ruck-sack in front of me to act as a buffer in this nightmare skaters' waltz. On every little fold and gradient, the children held me by the hand to keep me upright. It surprised me to see them walking without the slightest difficulty: they let themselves slide along with ease, keeping one leg turned so that the outer edge of the foot stayed at right angles to their direction. They left us at the bottom of the slope and ran away, sure-footed on the slippery rock.

Past the narrow entranceway, the flat-roofed house where Chris and his wife lived consisted of a single room with a curtain separating the bed from the 'sitting room', where I could see a table with a typewriter, three chairs, a stove, some pots and pans and crockery. The only luxury they seemed to have allowed themselves, apart from the linoleum on the floor, was the wallpaper covering the wooden walls. You couldn't see out through the window because of a coat of ice on the outside. In the sitting room were the two buckets that Joan used to bring water from the standpipe or ice for melting, according to the season. To supplement the very small allowance provided by their religious organization, Chris had taken a job as a packer at the store.

'If you need to go to the bathroom,' he told me when we came in, 'you know the custom of the country.' He pointed to the plastic bucket in the entranceway.

While we waited for the stove to get going, Joan made coffee to warm us up; we were drinking it when a Greenland woman wearing a blue anorak came in. Below her hood the skin of her face looked stretched by her high cheekbones, while her small, flattish nose seemed to start on a level with her brilliant, slanting eyes. She was very pretty. After saying hello, she took off her anorak; a beige V-necked pullover and tight-fitting ski pants showed off her beautiful figure. When she smiled she shook her head slightly and made her glossy black hair ripple down her back. A fringe over her forehead came right down to her eyes.

'Hello, Mitti,' said Chris and Joan. 'Sit down. Where's your husband?'

'He'll be here in a minute.'

Chris offered Mitti his chair and sat on the bed. She had just sat down when her husband arrived. He was the local photographer, a tall young Dane called Ib. Ib and Mitti were friends of Chris and Joan, and had dropped in to welcome them back. We had more coffee with the young couple, and when Ib learned that I had nowhere to stay as yet, he offered to put me up.

'But all we've got is a sleeping bag and the studio floor.'

I accepted gratefully.

Ib and Mitti lived at the other end of Jakobshavn, beside the frozen fjord, in a two-storey house, with another Greenland family living upstairs. Their apartment had two rooms and a modern toilet. Ib, who had installed their bed in the back

A view of Ilulissat

left-hand corner of their rectangular kitchen/living room, had transformed the bedroom into a darkroom, where we re-arranged his photographic equipment and materials to clear a space for my sleeping bag.

Next morning the first thing Ib and I did was to feed his dogs, throwing them fish heads from a bucket. In the afternoon Mitti took me with her to do some shopping and to introduce me to her parents. Instead of going along the main street, we took a short cut and scaled a rocky, slippery rise. At the top was a stretch of flat ground with a group of buildings including the church, an old people's home, and the new hospital, standing by the sea in a semi-circular bay. The Sistags, Mitti's parents, lived behind the old people's home. Next to the house was a lean-to with sledges hauled up on the roof. The family's dogs, thirty or so healthy-looking brutes, were kept in this pen made of wire netting and planks. Mitti's father welcomed us and told us the glad news that he had killed six seals that morning. The fresh skins, removed complete with flippers and claws, filled a wide basin in the living room. The wet fur seemed to be glued to the skins, which were very heavy because of the thick layer of blubber that had not yet been removed. The mother stopped work on them to make cof-fee for us. We drank it with Mitti's brother, a boy of fifteen, and her elder sister, Emma, whose appearance, haircut and manner made her look strangely like a boy.

Afterwards we went shopping at the store. On the way home, we passed the house where the explorer Knud Rasmussen had lived as a child; it was shut up and looked neglected. In the yard the icy wind was fluttering the last tatters of skin clinging to an old *umiaq* (the traditional sealskin boat for women) lying upside down on its supports.

This community was the most authentically Greenlandic of all those I had seen so far: there were plenty of kayaks, strong huskies, and sledges bigger than those at Sisimiut.* On an impressive number of drying racks hung fish, seal blubber, and fox skins. Though, with its 1,750 inhabitants, Jakobshavn possessed about twenty fishing and shrimp boats and a small cannery, this industry employed relatively few people. Most of the inhabitants lived by hunting, for the frozen fjord and its surroundings teemed with game. Seals were hunted and sometimes whale, especially the white whale. Apart from the large number of people living close to the harbour (where most of the Danish population was concentrated), the area near the frozen fjord was exclusively for seal hunters. Their houses, separated from Ib's by a stretch of rocky ground, were built of old planks and stood on peat foundations. These were the poorest dwellings, about fifteen in number.

Although there was game and wild life in plenty, dog meat made up an important part of the diet. According to Mitti, dog meat, when boiled for about four hours, became 'as tender as mutton'. She was backed up by Ib, who often ate it at his in-laws' house – in fact, they would be eating dog tomorrow. I learned that another family had some dog to eat, and because they hadn't yet skinned the animal, I went to their house to watch.

It was a very poor family, as you could tell by their belongings. I passed through the little entrance hall and entered the one narrow room which was home for six people, the parents and their four children. The furniture consisted of a huge doorless sideboard shoved against the wall behind the stove, which was to the left as you went in; one rickety chair; and lastly two shabby pallets fitted together at right angles, in the corner at the back. The four children were sitting on the threadbare

---

* The dimensions of a sledge on the northwest Greenland coast vary between two and three metres in length and are from seventy to seventy-five centimetres in breadth. The length of a kayak is 5.50 to 5.70 metres.

A hard lesson of handling a kayak

eiderdowns that covered the beds, playing among piles of old clothes. Their mother, very short, with a big head half buried under a mass of shining black hair, walked slowly up and down the room, smoking a cigarette, her right elbow cupped in her left hand which rested on a belly bulging inside her trousers. She had a vague, expressionless face. Her husband was out; she was waiting for his return because it was his job, she told me, to carve up the dog. To pass the time, I played with the children.

Finally Karl arrived. He took a knife, and we surrounded the dead dog which was lying on the floor. Killed three weeks ago and hung outside in the open air ever since, the frozen animal had now been lying by the stove for the last hour to thaw; but apart from the skin, which was now a little soft to the touch, its limbs were still rigid. Its jaws gaped in a mocking rictus; between its chops, above its blackish tongue, I could see the pale pink palate divided by a trace of black.

Karl crouched down, turned the dog on its back and began at

the left elbow joint, making an incision while his wife held onto the right foreleg to stop the carcass from rolling about. Then, thrusting the point of the knife into the slit under the fur, he slid it slowly upward as far as the chest, then across, and down again to the right elbow. As he cut the skin, he immediately separated it from the flesh with quick, deep stabs of the knife. The inside of the skin was greenish in colour. The excited children, looking forward to the raw liver and other tidbits still hidden beneath the flesh, kept jumping up and down round us and getting in their father's way. '*Tassa!*' he shouted, to keep them quiet. The animal's grinning head kept turning from side to side as they moved the body. The tip of the blade played the most important part; now it was gliding from the chest to the underbelly. A slip of the knife made a hole in the stomach, but nothing came out, as the contents were still frozen. Next Karl attacked the left side of the neck, cut under the jaw just below the eye, peeled off the ear, and scalped the head, revealing a very clean, shining cranium.

After whetting the knife blade on the stove, he slit the skin of the rear legs, one after the other, up to the groin, then continued these two oblique incisions upwards to join the slash down the underbelly, which looked more swollen now that it was bare. Neatly separating the skin from the flanks, spine, and rump, Karl removed the entire pelt in less than ten minutes, but left on the shaggy tail and the skin round the eyes and jaws and on the four drooping paws – that is, from the elbow joints to the claws. On the pinkish flesh you could see red spots and black sinews, but I never saw a drop of blood from the animal, which was still frozen solid. The most surprising thing was that, contrary to the usual custom by which only the women skin animals (especially seals), it seems that only men skin dogs, which are eaten raw or boiled.

It's the rabid dogs which are killed and eaten. The one which

Mitti's parents ate the next day was rabid, too. I don't know whether the dogs' long exposure to freezing air eliminates or reduces the risk run by those who eat them raw, but of all the mammals eaten by Greenlanders, only dogs are exposed in this way.

In my search for a smaller community where I could spend the polar night only among Greenlanders, on October 25 I took the advice of Mitti's father and made my first trip to Rodebay, a village of 170 people and 600 dogs, 23 kilometres north of Jakobshavn, facing Disko Island. The only Dane in the place was the schoolmaster. He was the first European settler and had been there only a year, teaching thirty-five pupils in a small chapel-classroom. Knud was his name, and he gave me a warm welcome.

'Leave Jakobshavn and come live here,' he told me. 'Then at least I'll have someone to argue with. I have a big house; you can have the upstairs room.'

Knud was thirty years old and unmarried. His solitude in a village cut off from the Danish colony was beginning to make him neurotic. The *palasi* (pastor), a Greenlander called Abili or Abilia (a corrupt form of Abel), helped him in his work. We spent part of the evening at the pastor's house, and he drank and swore like anyone else. As I was leaving, he told me, 'If you want her, I've got a cousin for you tonight.' Knud was shocked by this proposition. After visiting some of the inhabitants, all hunters, we returned to the schoolmaster's, where we talked and talked before going to bed around five o'clock in the morning.

I returned to Jakobshavn the next morning with the two young men who had taken me to Rodebay. On our way, we stopped to hunt eider duck. When we arrived that night, Mitti decided she would rather hang the two ducks I'd brought back and keep them until they were high. So she cleaned them, then brought in the liver and gizzards, which the two of us ate raw.

On November 6, at twilight, which came at four in the afternoon, we watched a whale being butchered on the shore in front of the church. A crowd of men formed on one side, a dog pack on the other. The white fur of the dogs, whose shaggy tails curved back over their hindquarters, formed a single seething mass, so close were they crowded on the slippery rock, fighting savagely over the scraps of meat that were thrown to them from time to time. The whale was tethered to the shore by a thick cable tied to an iron bar driven into the ground, and its massive head faced out to sea. First of all the men flayed one whole flank of the enormous creature, then they turned him over. They stripped off rectangular slabs of skin with a thick undercoat of blubber, then slabs of red meat. The long knives sank smoothly into the flesh; their users sharpened them by whetting one against another. A warm steam rose from the carcass. We ate some of the sweetish, tender skin (the *mattak*) on the spot. Mingling with the whale's warm blood, the thick layer of ice underfoot formed a reddish slush that made you slip on the rocky ground. Many of the men wielded the knives with their left hands. Expressionless faces turned to look at a team of huskies in front of the church, as they reared on their hind legs, jumped and howled, staring greedily at the free-roaming dogs gorging meat on the rocks. Their frantic yapping was fierce, sad and comical all at once.

In this early November, cold and blizzards took hold of Jakobshavn as the days grew shorter. The mornings were dull and sombre until eleven o'clock, when the no longer visible sun shed a feeble yellow glimmer on the distant island of Disko. Dimly, you could make out the icebergs standing motionless in the sea like great shapeless ghosts. Lights went on in the houses by four in the afternoon, and all you could see of the village

A pause on our way back from fishing through the ice
during the winter, Ilulissat

were the yellow squares of windows piercing the darkness here
and there. Though the carpet of snow made the landscape
blurred and slightly luminous, it was so dark that one day, at
four-thirty in the afternoon, I got caught in the traces of a husky
team which I hadn't seen coming; all I heard were the shouts of
the driver and the crack of his whip.

One night, the blizzard blew nonstop; a dull, monotonous
roar, broken with deep bellowing tones like the crashing of
breakers in a howling gale. Next morning, November 12, the
snow had gone! The warm, dry wind had swept it completely
away. There was nothing left on the rocky soil but patches of ice
and mud, while water now trickled down from slopes which
had been carpeted in deep snow only the day before. The sky
was blue again. It wasn't even cold any more – you could go out
in shirtsleeves. I was only lightly dressed when I went with Mitti
for dinner with Chris and Joan. But as we were leaving our

friends, around eleven, the weather suddenly changed: an icy wind started to blow, and the sea turned rough. When I got home I was freezing cold under my light shirt, with frozen ears and completely numb hands.

On clear days the moon gave us light from five in the afternoon, hanging very low in the sky and looking so extraordinarily large that it really frightened me the first time I saw it. It rose over Disko Island, but if you were coming from the other end of Jakobshavn, it seemed to be squatting on the rooftops in front of you, its crumpled silver-paper face splotched with grey. The play of moonlight on the icebergs was indescribably strange, and its magnificent refracted shimmers were brighter than day. One night, fooled by its brilliance, I got up at three in the morning. I soon realized my mistake, but the 'day' was so beautiful that I went walking for half an hour in the sleeping village. As it's seven-thirty in the morning in Paris when it's three-thirty in the morning in Greenland, I thought, in my calm and silent world, of all those Parisians with their steaming breath, now pouring into the Métro. I thought, too, about what a fine sunny day it must be back home in Africa.

While I was getting ready to go back to Rodebay, it occurred to me to get hold of some clothes made of animal skins; I mentioned it to Mitti's father.

'Dogskin trousers are good and warm,' he told me. 'I have a few skins I'll give to my wife to make up for you. Trousers of long white fur would suit you, and you already have a white anorak – Mitti can trim the edge of the hood with white dog fur. With two thick pullovers underneath, that'll keep you really warm. And you'll be really *kusanaq!*' he added.

He treated me rather like a son; in fact, the whole family already thought of me as one of them. Mitti's father and mother often joked about the big, strong, handsome boy I could make with one of their daughters.

Before my clothes could be made, Ib and his father-in-law were busy preparing for Mitti's twenty-third birthday – November 16, in just a few days' time. As there were no dry cleaners in Greenland, Ib had planned ahead and sent his woollen trousers and coat to Denmark to be cleaned, as all the Danes did. On the eve of the 16th, he went for a big dog-meat dinner with his in-laws. The following evening, the whole family and the neighbours filled our living room for the feast. Ib got drunk and fell asleep on my mattress just as table and chairs were being pushed aside for dancing. A rather too smoochy dance with Mitti had everybody watching us, and our upstairs neighbours congratulated her on having made a conquest of me. Her mother, flattered, beamed with happiness, and her father looked at me pointedly, saying:

'She's pretty, my daughter, isn't she?'

'There's no one to match her!'

'You like her?'

'Who wouldn't?'

The last guests, our neighbours, left at four in the morning. Mitti's father, when he got up to go home with his wife, hung back a moment to let me know – with a wink in Mitti's direction – that he intended to ask Ib to go hunting with him for a few days.

'We'll trek across the frozen fjord and camp in tents on the edge of the ice.'

Sure enough, he asked Ib over to his house to talk about it, and on Thursday, the 18th, he set off with his son-in-law at eight in the morning, so as to leave me alone with his daughter.

'My father likes you very much, and so do I,' Mitti told me that night, as she lay down next to me on my mattress on the floor.

Her father and Ib were away four days. For Mitti and me, those were crowded days. From that time on, Mitti kept giving me little gifts: sweets, chocolate, packets of coloured candles, and two candlesticks to hold them.

Night was now falling at three in the afternoon. We wouldn't see a ray of sunshine for months on end. I was oppressed by this sombre night that had steadily lengthened and now reigned supreme. If I had come to northern Greenland straight from Africa, I think that this total absence of sun month after month would have driven me mad; it was lucky for me that I'd been able to live through the slow onset of the polar night in easy stages.

The sea was not all that frozen yet. Only a crust of whitish, rock-stained ice floated along the shore. The hunters from the poor end of town, by the edge of the frozen fjord, redoubled their efforts to kill as many seals as possible before the sea froze over. From then on I spent most of my time with them, helping them after the hunt. Sometimes five or six kayaks returned side by side, covered with a thick sheet of ice formed by sea spray on contact with the icy air. We would drag the kayaks onto the rock, then remove the coating of ice by scraping or chipping it off with a knife or simply a piece of wood. Each hunter came back every day with two or three seals, four at the most, and a few birds – eiders, ptarmigans or guillemots.

After collecting the birds, guns and harpoons, and the kayak paddles that the hunter sometimes carried over his shoulder, we made our way home, hauling the seals after us. To do this, you pass a double leather thong through the eyes and tie it in front of the snout. This enlarges the nose hideously. The dogs always pounced upon the animal, biting it and licking traces of blood off the ice-sheeted rocky soil. To prevent gravel from damaging the sealskin when crossing the street, we placed the animal on a sheet of corrugated iron abandoned by the side of the road. If we couldn't find a sheet of metal or cardboard thick enough, and the seal was too big to be carried, we passed a belt under its two flippers, then one of us hoisted it onto his back, hanging on to the two ends of the belt where they passed over his head from nape to forehead, over the hood of his anorak. The seal hung down his back to the ground, its head flopping back. The kayak was carried either by one man holding it against his hip with one arm inside the opening, or by two men each holding one of the pointed ends. Outside the hunter's house we made it fast either on supports or on a drying frame, well out of the reach of the dogs, which eat anything made of animal hide, even leather whips. A slip knot enables the hunter to get his kayak quickly and carry it down to the water whenever a sea mammal is sighted.

One day five fishermen in dogskin trousers – men who lived all together in one of the houses by the fjord – arrived in a boat with two seals, one of them very big, with a white belly and only two small black patches on its back: a fine skin! The animal had been killed by a rifle, so there was a big bullet wound in the neck. I helped the fishermen drag the catch up to their house.

This consisted of two rooms, each with a bed piled with clothes. There was only one woman and two children. I was given a chair – the only one in the house – and the fishermen removed their clothes. They looked younger, not so fat, and strangely taller than when wearing their fur trousers. They rolled these up on the floor and left them there beside their boots, then went into the next room. The two seals were placed on sheets of cardboard on the floor.

The woman squatted down to carve up one of the two seals. With her *ulu* she first made a cut from the belly to the head, just under the jaw. All along this slit, the yellow blubber underneath the skin opened like two enormous lips. The skin, together with quite a thick layer of fat, was cut loose first on one side, then on the other. In places we could see the darkish flesh appear. The woman broke the two forelimbs, which she left with the skin. The skinless, blubbery body lay on the cushion of fat attached to the skin now spread out beneath the seal. Blood ran across the cardboard and overflowed onto the floor. A quick thrust of the *ulu* into the stomach, and the entrails were emptied into a bucket. Then she made several incisions along the intestines, which she emptied of their contents by squeezing them between her fingers and pulling with the other hand. The lungs were cut into slices and eaten raw by the children, who, before putting the pieces in their mouths, dunked them in blubber. Their hands, lips and cheeks were daubed with blood. The embryonic, flipper-shaped hind limbs were left unskinned on the animal. Then, squeezing with both hands, the woman drew the blood from the belly and poured it into another bucket. After that she cleaned out the stomach with a rag; the blood-soaked cloth was wrung out over the bucket so as not to lose the least drop. The woman herself was drenched in blood: she held out her arms to her husband – one of the five fishermen who lived there – and he rolled her sleeves up to the shoulders. The *ulu* was whetted against another knife. The woman broke the seal's ribs. The flesh, cut out piece by piece (perhaps according to a strict procedure), filled a large container. On the sheets of cardboard, the fat was red with blood. Now, from one end to the other, the woman loosened the skin from the fat, and the fat rolled off, thicker than melon pulp in its rind. This fat was chopped up and thrown over the fence to the dogs' enclosure. You should have seen them pounce on it – what a wild rush! As I left the house, a pack of dogs was clustered around the doorway, ears pricked, eyes blazing, waiting their turn.

Later, going with Mitti to her parents' house, I found on the living-room floor a tiny little baby seal taken out of the belly of its dead mother. Of course it was no longer alive. It had a pink body, redder forelimbs, fine, slightly shiny fur, with darker ringlets developing under the skin, a triangular mouth, and toothless gums. The umbilical cord still dangled. It had no tongue, as if the tongue were stuck to the palate. Two minuscule holes instead of gills, and above the unopened eyes there were two patches. The massacre of baby seals, culled shortly after birth, does not take place in Greenland; the presence of this little beast on the living-room floor was a sheer accident.

In his impatience to get me to Rodebay, Knud the schoolmaster sent a Greenlander to fetch me in his boat. I had no plans to leave on that particular day and my dogskin clothes weren't ready, so I had to send the man back with a letter in reply to the one sent by Knud. Meanwhile I kept urging Mitti's mother to finish my trousers, but she didn't think she'd be done in less than a month, by which time the sea would be frozen over . . .

Two days later I visited Karl, who took me for coffee at the house of one of his neighbours, Aqqaluk, whose wife had just made him a pair of white dogskin trousers which were too long for him. I tried them on and they were a good fit, so I bought them and a pair of *kamiks* for 250 kroner.

'It's too dear,' said all the members of Mitti's family, and off went father, mother, Mitti and her brothers and sisters to have it out with Aqqaluk. When you're adopted by a family, they not only try to make a place for you in the community, but they also do all they can to defend you and to keep you from getting rooked.

Aqqaluk eventually agreed to come down to 120 kroner, which everyone said was fair. Mitti trimmed the edge of my anorak hood with white fur. For 20 kroner Karl's wife made me a pair of mitts faced with long-haired black dog fur and lined with sealskin, supple enough so that I could just about clench my fists.

When I went out for a walk in my fur clothes, there was general admiration, although to me they felt too thick and seemed to triple my bulk. Because of the long hairs on the trousers, I had to walk slightly bow-legged. But on the whole I felt comfortable in them: the flat soles of the *kamiks* gave a better grip on the rocky road.

The next day – November 30 – I went to the harbour to catch the small boat owned by the Danish trading company that supplied the village of Rodebay. Mitti saw me off. Without looking at me, she squeezed my hand. Our farewells were calm and simple, without sadness or tears.

'Be seeing you, Mitti.'

'Yes, see you,' she said, after buttoning up my overcoat. 'You'll be seeing me in Rodebay. I'll come and visit you. Write to me.'

We looked at each other one last time, then I went aboard. Apart from the pilot, a big Greenlander named Evat, and his two mates, I was the only passenger.

3

My Host, Thue

We set off at ten o'clock, navigating in total darkness by the light of the masthead lamp, and arrived an hour and a half later at Rodebay. The twenty or so houses that make up this village are built on a gentle rise behind the tranquil bay, which Dutch whalers in the last century called 'Roo Bay', or 'Bay of Rest', which gradually became Rodebay. In Eskimo the village is called Oqaatsut, meaning cormorants.

Three young men helped me carry my luggage to the school-master's house. Knud rushed out like a madman and refused categorically – even brutally – to take me in. He even ordered me to go straight back to Jakobshavn, as if Rodebay belonged to him. Faced with this attitude, which struck me as odd, to say the least, I went to the village headman, Johan Dorf. He was surprisingly off-hand. Next I hurried to the pastor, who refused to see me, pleading a sudden illness. His wife's greeting was as cold as could be. Yet these were the very people who had received me so warmly just over a month ago, and this same pastor had suggested I might spend the night with his cousin! All the inhabitants shut their doors against me and unanimously refused to welcome me to Rodebay. No door opened to my knock. I felt I must be dreaming . . . For a whole hour I tramped vainly from door to door in the dismal cold and over ground thick with ice: my feet went numb in my *kamiks*. What could have changed these people so completely in so short a time? I soon realized that here the word of the little Danish schoolmaster was law.

Upset by this shameful treatment, which the whole village later blamed on Knud, a young man named Saqqaq (a local distortion of Zachariah) – one of those who had helped carry my luggage from the ship – came up as I sat on my rucksack in the street and invited me back to his house. He took me to the other side of the village, to a tiny house which was the most isolated in the community, and the most ramshackle place I'd ever seen. As I went in, the occupants crouching in the living room looked up at me with a strange, bright stare. A girl named Maria, aged thirteen, pulled off my *kamiks* and sat me by the stove to get warm, while her elder sister Marianna stowed my bags in a corner. That's how I came to stay with the Thue family (pronounced Too-ay). Thue was the poorest man in Rodebay, and I soon learned that he was the worst hunter and the most hated man in the village.

Eleven people lived in this tumbledown shack: Thue Petersen; six children (four small boys and two daughters; another daughter lived in Egedesminde); his nephew Saqqaq; his son-in-law Hendrik Olsen, husband of Marianna, and their two children. Thue's wife had TB and was in the hospital at Godthåb. Since Maria was still going to school and Marianna, the eldest daughter, took more care of her husband and children than of her father, there was, to all practical purposes, no woman in the house. Thue, a sensitive man, lived in deep solitude among his own offspring.

I took in the contents of the living room with a single glance. A coal bucket stood next to the stove; another bucket over the fire was filled with lumps of ice being melted for water. Just above the fireplace hung a wooden rack where boots, mitts and socks were drying. On the filthy floor lay a few empty powdered-milk cans used for drinking water or coffee, and next to them a knife. A third bucket, made of plastic and intended for calls of nature, was placed behind the door, next to a little table which held a white bowl, yellowed on the inside, for morning washing.

In the bedroom, where I soon went to put away some of my things, there was a table with a black transistor radio on it, and two beds for the entire family. Two dilapidated suitcases with rusty, twisted locks stuck out from under the right-hand bed; under the bed on the left were two Tuborg beer crates. These old boxes and suitcases contained all the family's possessions, apart from Thue's hunting gear.

There was no food in the house, no store of frozen meat outside; the whole family – like their few scrawny dogs, which had no enclosure – lived in the shadow of starvation. They had the withdrawn look produced by hunger pangs.

When he heard what house I had been taken to, Knud changed his mind. Late that afternoon, in the space of only a few minutes, he sent me two letters through Pavia, his Greenland houseboy. 'The family that has taken you in is the poorest in these parts and has nothing, absolutely nothing to eat,' he wrote. 'I beg you to come to my place, if you don't want to die of hunger!'

'Tell your boss I mean to stay where I am, and don't you come back a third time!' I told Pavia as I sent him away.

In any case, Thue didn't want me to leave. Possibly he thought he'd stumbled upon a gold mine.

'Don't go to Knud's,' he kept telling me. '*Qallunaaq ajorpoq*.' (That white man is no good.)

My stay in that house was to show me once more the crying lack of mutual help in a Greenland village, and the villagers' profound contempt for their poorer countrymen. Thue was not an invalid and he was under sixty, so all he received were small allowances for five of his children (Marianna and the daughter in Egedesminde were over seventeen), amounting to about fifty-five kroner a month in all. With this money he bought the coffee and powdered milk that his family mixed with warm water and drank as their dinner.

In the evening, after this meal, I went out for a breath of air. No sooner had I gone down the three front steps than one of the ravenous dogs sprang at me. All I felt was a swift, violent blow on my thigh, accompanied by a loud snapping of jaws and a growl. Fortunately the fangs couldn't penetrate my thick dog-skin trousers, under which I wore two old pairs of woollen trousers as well. Before I could make out which one it was in the darkness, the animal turned tail and disappeared beneath the house. The dogs had made their home in the glacial crawl space between the frozen earth and the floor.

They put a mattress for me in the living room, which was a little warmer than the bedroom. Marianna, her husband Hendrik, and their two children slept in the big bed on the left; Maria and two of her younger brothers had the second bed, while Saqqaq, Thue and Thue's other children slept on the floor. They kept the door open between the two rooms so as to get a little warmth from the living room before the stove went out.

I spent the most terrible night there. From the other room, I could hear Thue talking through his nose in his sleep, as clearly as if he had been next to me. Sometimes he raised himself on his elbows, his eyes still closed and his head in his hands, and went on soliloquizing, rocking gently backwards and forwards all the while. The children on the ground also sat up from time to time, then let their heads fall with a thud against the floor – all without waking up. One of the boys kept crying, and Marianna's baby never stopped whimpering: she didn't bother to get up to see to it. Over my head, the icy wind blew through a broken windowpane. I got up to put on all my pullovers, as well as my overcoat and three pairs of trousers.

Thue rose at seven. He did the same things and went through exactly the same routine every day, with rigorous precision. Clad in an old checked shirt, open over a filthy undershirt, and wearing cotton long johns stained with big spots of yellow, he left the bedroom and went to piss in the bucket, all the while scratching, wheezing and coughing. After that, he started looking for his anorak and *kamiks*. '*Naak kamikka*, Maria?' he would shout. I raised my head; with a beaming smile he motioned me to lie down again, but by now I was no longer sleepy, so I got up, too.

The *kamiks* were right there under his nose; but it had to be a woman who came and took them off the drying rack and got them ready for him by putting the sealskin socks in the boots. So little Maria got up, her eyes still heavy with sleep, and came into the living room to pick up the *kamiks*. Before getting them ready, she lit a good fire and put the bucket half-full of ice on the stove. The ice began to crackle. She added more blocks of ice to the water produced, and they gave off the same dry little cracks until they were completely melted. Then she put on some water for coffee. While Thue drank his coffee and the others got up one by one to come and crouch by the fire, little Maria had to go out into the freezing cold to collect the lichens used for stuffing her father's *kamiks*. She found them in places not yet completely covered by snow, or else had to scrape the snow away from the ground. These lichens she sorted and pressed down into the boots to form an insulating layer between the sole and the fur socks.

We clustered round the stove. By now we had all had our coffee. Saqqaq lit his pipe. Maria, sitting in profile in front of me, rocked her sister's baby on her knees for a while. Then she put the baby back in the bed and started to sweep the living-room floor with a seagull's wing. Once white, this wing was now a sordid, dirty grey. The sweepings were not thrown away but piled up by the coal bucket to be used for kindling. Then the little girl washed herself and went off to school.

His daughter's departure seemed to remind Thue that he had to go hunting, and he started to get ready. First, he went and sat on the privy bucket. This performance, which we all went through one by one after coffee, and which in the mornings was carried out in front of the family, took on even greater importance during the day and especially in the evening, when there were visitors. You sat in the living room with your guests. When someone wanted to relieve himself, he got up quietly and made for the bucket. When he got there, he turned and faced the company again. A man would drop his pants or a woman lift her dress with no false modesty, then squat on the edge of the receptacle, all the while following the conversation and talking to the others.

Finally Thue got dressed: over his long drawers he wore ski pants with understraps, then a pair of thick, black trousers, the backside shiny with streaks of dried snot. Over his undershirt he put on a single pullover and a green cotton anorak. Then he pulled on his *kamiks* and lined the sleeves of his anorak with pink rubber half-sleeves that reached from wrist to elbow. He threw back the hood to pull on a sort of green cloth cap, padded and quilted inside and on the peak, with earflaps ending in strings that tied under the chin.

He went out to get a big lump of fat from its storage place on the roof, and a *tuiitsoq*, a kayak apron made from seal-skin, rather like breeches without a crotch, ending in a hem threaded with a leather thong. Before sliding down into the circular opening of his kayak, the hunter puts this apron on above his trousers, like a short skirt; then, sitting with his legs stretched out in the kayak, he ties the bottom of the *tuiitsoq* around the edge of the hole, which makes him and his craft into a single unit and also keeps it watertight.

The apron had been outside all night and was as stiff as a sheet of iron. Thue flexed it vigorously to make it supple, and rubbed it with thick yellowish blubber tinged with thin streaks of blood. Then he hung it from a bent nail in the ceiling; black and gleaming with grease, the skin was transparent in places, probably where it had been scraped the hardest during its preparation. Thue warmed his mitts, stuffed them with the last of the lichens, which he crumpled up between his palms, then unhooked and warmed the apron, and went to launch his kayak.

His hunts lasted for only two hours a day. In fact, it was only eleven o'clock when, standing on the shore, I saw him already returning, alternately dipping his paddle to left and right without stopping. He paddled up to the rocky shore, removed his by now softened apron, then heaved himself out of the kayak and clambered onto the rock. Entering and leaving this fragile craft is a difficult operation that requires careful balance. Thue unloaded his rifle and put away the long paddle with its flattened blades tipped with polished white bone. The kayak, carried one-handed against the hip, was tied onto two protruding beams of the big drying rack outside the house. Everything, even his mitts, was left outside.

Thue had brought back no game, not even a bird, so he asked me for money to buy coffee, biscuits and some *puuluki* (bacon). Instead of all that, however, he bought ten bottles of beer. Just as he was entering the house with the carton clasped to his stomach, one of his own dogs attacked him, and two of the bottles were smashed. He asked me for three more kroner, then for another three and fourteen øre so he could send for two replacements. Naturally I refused, and our relationship cooled temporarily.

In the afternoon, giving in to the pleas of Knud the

schoolmaster, who blamed his outlandish behaviour on his nerves, I went to pay him a visit. When he left me to go back to class, his neighbour Cecilia, also regretting the way she and her husband Hans had treated me, came to invite me for coffee at their house, where, because of the cold outside, she warmed the skin socks of my *kamiks*. She served me a cup of coffee, poured beer for Hans, and gave me some seal meat to eat. Cecilia had prepared the lungs of the seal for her husband, who loved them. A short woman with a coppery complexion, a long face with curious vertical wrinkles, and grizzled long black hair, she looked like an Amazonian Indian. She showed me the fresh skin of a seal her husband had killed that day. Contrary to what my reading had led me to expect, Hans had not shared his seal with his neighbours; which is why this household had food in abundance, while Thue's family was starving nearby.

Even the dogs of the other villagers didn't go short of meat. Every day the Thue family and I watched through the window as a huge pack of dogs on the frozen sea devoured long slices of *eqalussuaq* (blue shark), which we ourselves wouldn't have scorned right then.

Seeing her father come home empty-handed as usual, Marianna complained. In the evening, Maria came home from school; she didn't add her voice to her elder sister's grumbles, but kept her feelings hidden, except when she was happy. She was an ideal daughter, quite unlike her sister, who spoke in a monotone, baring her teeth like an angry dog. Maria went out and stood on the doorstep, unaware of being watched. She was silhouetted in black against the grey background of the sky, her profile precisely outlined against the rosy streak of the horizon, her long hair falling to her shoulders. For a few seconds she stood staring vaguely into the distance. Whatever she felt at the moment about spending another long night without food, she kept to herself. Then she came back in, smiling, called the

children, and started playing with them, from time to time casting an affectionate glance at her father.

Thue came back empty-handed twelve days in a row. On
the thirteenth, he killed his first game, a scraggy *malumuk*.* The
bird was prepared at once by Maria, who neatly removed the
skin, feathers and all. The wings, broken off at the shoulders,
would be used as brooms. But what use was one bird to twelve
starving people? For some reason that I didn't understand,
Saqqaq and Hendrik never went hunting. Perhaps it was because
there was only one kayak and one old rifle.

Things being as they were, Thue began to slaughter his dogs
one by one to feed his household. When he was about to kill
the third dog, I stopped him.

'I'll see if I can borrow some money from Knud,' I told him.

So, for about two hundred kroner, I bought seal meat from
the neighbours, rice at the store, and some powdered milk
reserved for Marianna's baby – provisions for a whole week, in
the hope (never to be realized) that Thue would come home
one day with a seal.

A thick coat of frost made our three windows opaque. To
look through the frosted panes at a little patch of milky-white
landscape, dotted here and there with black specks of men and
dogs in the snow, we breathed on the ice to melt just enough
space for an eye. The only way to prevent, or at least delay, the
formation of frost is to double-glaze the windows. Thue
couldn't afford that, and when the house warmed up a little, to
fifteen degrees, the frost melted and ran along the frames and

---

* Its name comes from *malik*, waves, and *mut*, a preposition indicating direction (going towards). So it means: 'that which brushes the waves from time to
time when gliding'. *Taateraarnaq* means seagull: 'that which flies smoothly,
always at the same height'. *Appa* is the guillemot: 'that which flies with many
wing beats'. Similar to a small penguin, the guillemot is white and black; it is
called *alk* in Danish. There are, however, no penguins in the Arctic.

water gathered on the windowsills, soaked into the woodwork, rotted the supports and cross-pieces, then seeped along the wall and flooded the floor. Everything inside was a squalid, unsanitary mess.

At every moment of the day and night, all over the house, the children would shit or piss where they stood or sat, especially now that they were eating better. Like the others, Thue used thumb and forefinger to pick his nose and drop snot into the coal bucket or the pisspail (which doubled as a spittoon), then wiped his fingers on the seat of his pants.

Aqqaluk, who was eleven, got down on all fours in the bedroom and sucked the prick of his two-year-old brother Anganngut as he stood on the bed, wearing only a shirt. The boy's shivering and the sight of his erect member made everybody laugh. Aqqaluk repeated this game two or three times a day without being reprimanded by anyone.

Except for Johan Dorf, the village headman, who still wore a dark, cross-tempered look whenever we met, the other villagers, even the pastor and his wife, became more friendly and helpful. I was glad to see this welcome change come over the people with whom I was to spend the long winter months, cut off from the rest of the world by the ice already covering the sea.

During the previous two weeks, ice had started to build up each day, but each time a blizzard shattered it. Once again the waves smashed hard against the rocky shore in showers of spray. Slabs of broken ice bobbed on the waves like great white water lilies.

That was my first sight of the sea freezing over, and I kept a keen watch on this furious battle of cold and waves and ice. It was the same fascination I had felt in my childhood when, hidden behind a tree, I'd watch a battle to the death between two snakes.

Eventually the surface of the bay was nothing but a vast white stretch of pavement strewn with numerous black or blue patches that made it look rather like marble. When first I walked out on the frozen sea, it gave me an unforgettable sensation, at once pleasant and frightening. While others strode out firmly, I planted my feet with care. I was scared but refused to show it. What if the ice, which was not supported by anything underneath, should suddenly break? Just now it was perhaps less than one metre thick, and a mass of invisible water flowed underfoot. I saw myself sinking into the freezing water with no hope of getting out, because of the immense sheet of ice that stretched above my head like a ceiling: I would drown in no time. My foreign upbringing may have exaggerated these fears, but they were not imaginary: a few Greenlanders die this way every year, drowned with their dog teams when the ice breaks beneath their weight. A particularly strong individual who manages to get out of the water can still die either by freezing (when the water on his body turns to ice on contact with the air), or by snapping his frozen spine when he bends down. My mind buzzed with the stories I had read or heard told here about such accidents.

It was two o'clock in the afternoon – impossible to see anything without a light. Villagers with lanterns were busy on the ice: great icebergs drifting near the coast were now imprisoned by the freeze; using an ice knife (a sort of chisel with a long wooden handle), they broke off big lumps which they loaded on their sledges to be melted down at home for drinking water.

Out at sea, the temperature of the frozen ocean rises slightly, deeper down, and the current continues to erode the base of the imprisoned icebergs. Sometimes, after weeks of standstill, the upper part is left heavier than the eroded base, and the icebergs capsize, enormous masses turning over on themselves

Breaking off big lumps of ice we used to melt down
for drinking water, Oqaatsut (Rodebay)

with a dull roar and cracking the ice for several hundred metres
around.

On Monday, December 13, the Jakobshavn trading company
boat made its last but one visit to Rodebay for the winter. Evat,
the Greenlander pilot, informed me that a helicopter had
arrived in Jakobshavn just before he sailed, but that he hadn't
been able to wait for the mail. I thought the money order I'd
been waiting for since the beginning of the month might be
among those letters, so I returned with the boat to Jakobshavn.

In December the helicopter bringing the mail from Godthåb
to Jakobshavn flew in only once a week, on Saturdays. Because
of bad weather, this weekly flight was often postponed until the
following Saturday. Sometimes we went without mail for two
or three weeks, even a month. The next helicopter then arrived
on some day other than a Saturday, as was the case today. If I

didn't receive the money order now, I wouldn't get it until March or April, with the arrival of the first boat of the year in Rodebay. If Christmas presents from Denmark to a northern village like Rodebay aren't sent by the beginning of November, the precious parcels won't arrive until towards the middle of the following year when the ice melts.

Already, on that Monday, December 13, a tall illuminated Christmas tree ornamented the front of the store in Jakobshavn. At the post office I collected a letter from my adoptive father, dated late November, saying that he had sent the money order from Paris the day before. The clerk emptied the mail-bags on the floor, sifted through the letters and parcels, then looked up and told me:

'I'm sorry, but your money order hasn't come yet.'

Understanding my difficulties, the manager of the trading company, who among other things acted as postmaster, advanced me the six hundred kroner, and I returned to Rodebay the next day in the company's boat.

The boat was loaded with fruit, mainly apples, intended for Rodebay and for Saqqaq, another village further north. A young company employee, a Dane named Ole Würtz, came along to deliver these goods to the Danish stores at these two villages. Ole and I had to sit on top of the stacked boxes of fruit, an awkward perch made all the more uncomfortable by the rough sea. The water flooding across the decks froze solid immediately, so that by the time we arrived the bridge and hull were covered with a thick crust of ice which had to be chipped off with a shovel. The boxes of fruit were unloaded onto the frozen fjord, then transported by dog sled to the village.

The villagers greeted me with friendly smiles and heaved a sigh of pleasure: 'Ah!' Thue's children were waiting for me, standing on the ice. I disembarked and we went home on foot. Already I was much surer of myself on the ice, even in places where it was not covered with snow.

Thue at once asked me for money to go and buy a case of twenty-four bottles of beer.

'Impossible,' I told him. 'When I've paid Knud back what I owe him, we'll have only four hundred kroner left to last for several months, and that's not much. Let's keep this money for food. The bay is frozen over and soon I'll be able to help you fish through the ice.'

But Thue didn't see things that way. In moments like this, his kind of village drunk turned nasty, so I had to be tactful. I went out to buy seal meat for all the family. When I got back, since I'd eaten nothing all day, I was brought a plateful of some kind of meat, an astonishing bright red in colour.

'*Suumuna neqaa?*' (What kind of meat is this?)

'*Qimmip neqaa!*' (Dog!)

Thue had killed yet another of the household dogs during my trip to Jakobshavn. Its bones lay in a basin near the stove.

I shoved the dish away. In spite of my numerous opportunities to taste it, I could not overcome a deep aversion for eating dog. Likewise, Greenlanders would probably be put off if they had to eat the unclean beasts enjoyed by some African tribes – monkey, crocodile, or snake (which I personally have never been able to eat, regardless of any religious considerations). And their repugnance might even be greater than my own for cooked dog, simply because they aren't familiar with these animals from our world.

The children made short work of the dog meat. Hendrik suddenly came in, wolfed down a portion of raw dog meat, and left the bone on his plate, furrowed with small blood vessels.

While the seal meat was boiling (in one of the unwashed pans used for the dog meat!), I went over to Thue, who was sulking in a corner. He wouldn't eat if I did not buy beer and *akvavit*. Finally I gave him fifty kroner.

'I want another hundred!' he yelled. 'You owe me money!'

'I owe you money? How do you make that out?'

'You ought to pay for living in my house.'

He had me there!

'You're a bad *Qallunaaq*!' he shouted.

I still refused to give him the other hundred kroner, because he was bound to spend them on drink. And I wasn't mistaken: on the strength of the fifty kroner I had given him, Thue invited several people to come for a drink, including one old man who got drunk and fell flat on his face. Sitting there on the floor, Thue and the rest took no notice, but went on drinking and listening to a record request programme on the transistor. Aqqaluk went out three times to buy more booze, and by the end of the afternoon Thue didn't have a single kroner left as he lay bleary-eyed in the middle of the living-room floor, not knowing what he was rambling on about.

In the evening Hans and Cecilia invited me for coffee and, knowing what a bad landlord Thue was, they suggested I stay with them. Hans was a good hunter: besides his kayak, he owned two good rifles, fishing nets, two sleds and twenty-four healthy dogs. Eager to learn to hunt from a man like him, five days later I moved out of Thue's house, helped by Saqqaq and Poyo.

Hans and Cecilia had no children of their own, but they did have two adopted sons. Poyo, aged nineteen, spent his time fishing through the ice. The other son, Izâ, who was seventeen, worked in Jakobshavn on a factory shrimp boat and so didn't live in Rodebay.

Two days later the whole village was in an uproar. During a furious blizzard, a twisting channel about one metre wide and almost a kilometre long had opened in the ice from the bay right out to sea. A fracture like this allowed air and some dim light to filter into the dark depths, and this attracted seals. The excited villagers were hoping to net the seals that would come up there for air, before it froze over again.

Hans took a coil of rope and one of his *qassutit* – rectangular nets, usually blue in colour. I carried the lamp, and we went down to the bay. With his ice chisel, Hans dug a hole, moved four paces away and dug a second hole, then a third, all in a line along the narrow channel. They were just over three metres apart, and the space between the first and the last was equal to the length of the net, or about seven metres. The idea was to suspend this coarse-meshed net vertically in the sea beneath the ice, holding it in place with three cords running into each of the holes. The net was not baited: it was simply a sort of snare for the seals, which are short-sighted, as they tried to cross the channel at this point. The animal sticks its head into the mesh right up to the shoulders, and even to the belly. In its efforts to wriggle loose, it pushes its head through several more holes and ends up completely entangled. Then, no longer able to rise to the surface to breathe, it suffocates. The hunter, who may come back only every two days to inspect the net, drags it out and replaces the 'snare'.

Sometimes, though seldom, two or three seals get caught in the same net. A hunter may own as many as ten *qassutit* and set them in different channels, often several kilometres from the village. To make the net hang plumb and to stop the water pressure from thrusting it up to the surface, the edges were weighted with light stones at regular intervals. Sometimes, as I was to see later in Christianshåb, old car doors are hung at each end of the net and do just as good a job.

After digging the holes and tying on the stones collected from the shore, we stretched the net out between the two of us and let it down into the water, each holding one of the ropes. While I stood beside the first hole and kept one end of the net out of the water, Hans lay full-length on the ice beside the last hole with one sleeve of his anorak rolled up, then passed his rope under the ice, pulled it up through the hole, and tied it to a block of ice placed in front of the opening. Although his hands were bare and numbed with cold, he took his time securing the other two ropes in the same way.

'*Ajunngilaq, takuuk!*' (That's fine, take a look!)

I bent down to look through one of the holes: we'd done a good job! Through the calm, clear water I could see our net nicely stretched, hanging down to the bottom, its upper edge just touching the floor of the ice-floe. We collected the ice chisel, the rest of the rope and the lamp, and went home.

Next morning we went back, equipped with some net cord and a netting needle (sometimes the seals damage the net, and it has to be mended on the spot), an electric torch, the ice chisel and a pocket mirror.

'What are we going to do with the mirror?'

'You ask too many questions, Mikili! Wait and see.'

The channel had frozen over again, but the net's location was marked by the thinner ice and the three ice blocks. Hans dug a hole. We bent down and looked in: the water was clear and there was no sign of seals, but we couldn't see all of the net through this one hole. Hans dug another hole seven paces further on.

'Stay where you are,' he told me, 'and shine the light down there!'

He went back to the first hole and lay flat on the ice, holding the pocket mirror in the opening. By turning it one way and then another, he could inspect the whole length of the net.

'*Puiseqannilaq!*' (No seal!)

We got one that evening. The dead seal was rolled up in the net, lying against the floor of the ice floe. Hans made a hole right over the animal, and we enlarged it as we pulled it out of the ice.

We captured five or six seals a week in this way. They were nearly all *Phoca hispada* seals (*natseq* in Greenlandic), which measure between 1 metre and 1.60 metres long, and are called stellated or annulated 'fjord seals' because of the ring-shaped markings on their fur. Their skins are used for making men's clothes, especially hunters' trousers, or, with the hair removed, for covering kayaks. Old kayak skins are often used to make harnesses for dog sled teams. In the old days they were also used to make tents. Sold to the trading company, a stellated sealskin undamaged by harpoon or rifle bullet (and cured by the women, who stretched it on a wooden frame to make it flat and smooth) could fetch thirty-three kroner. Up to 52,000 stellated seals are killed each year. Their flesh is the most tender of any species. It also happens to be the most prized by polar bears, which catch seals by putting their great paws through the ice; from them man learned to hunt at breathing holes.

Stellated seals do not migrate in winter. These mammals can remain submerged for twenty minutes, but surface to breathe at regular intervals of seven to nine minutes, so they make breathing holes through the young ice as soon as it forms on the sea. As the ice thickens, they keep the holes open with their claws, their teeth, and their warm breath. Holes made in this way eventually look like chimneys through two metres of ice. They are called *allu*, seals' breathing holes, and are masked by the snow.

To catch seals at the *allu* – a kind of hunting in which a harpoon with a detachable head is used – it is the dogs which sniff out the breathing holes hidden under the snow. But during my stay in Rodebay, none of the inhabitants tried this kind of hunting, which requires meticulous preparation and immense endurance. In front of each *allu* the hunter carefully places a marker, which can be either a very sharp piece of bone tied to another and fixed in the snow by a string, or one or two eiderdown feathers attached to a little piece of frayed tendon hanging freely in the opening. A slight movement of the marker warns the hunter that the seal is surfacing to breathe, and he must harpoon it through the ice with all his might, throwing blind. Hit in its muzzle or neck, the seal races furiously away with the detached harpoon head embedded in its body and tied to a long leather thong which the hunter uses to pull the seal out of the ice, when it has exhausted itself under the water.

As each seal has several breathing holes, you have to bide your time until it comes up for air at your hole. There are stories of hunters waiting for a kill like this for two days, sitting on a stool on the ice. The seal has acute hearing and is very cautious. So using the net has one great advantage: you can leave it stretched under the ice and go about your business.

Seal blubber sold at twenty-four øre a kilo, but Hans preferred to keep it as food for both ourselves and the dogs.

One day, as we were returning home with one of these seals, I asked timidly if we might not share it with Thue.

'Can't he hunt for himself?' he snapped.

We never mentioned it again. I was now convinced that I was living among people no different from any other men on this earth, and that their communal spirit had simply been exaggerated.

Hans and Cecilia took me to dinner with Augustina and her husband Jørgensen, their neighbours and friends. When we reached the house no meal was ready, but a whole seal, caught by netting, was waiting for us on the floor. The animal was placed on a sheet of cardboard, its fore fins resting on its plump belly, the lower part of the stomach slit open and the intestines spilling out. As soon as we had sat down at the table, Cecilia got up, took a bowl which Augustina handed her, and emptied the intestines into it by pulling them out hand over hand. Slowly the belly subsided. Augustina used a knife to slit the seal further towards the chest: the layer of blubber that then emerged was yellower, not so pink as the layer already exposed. She slid her hand into the stomach, delved expertly using only her sense of touch, and tore out bits of the lungs and then the liver, which we ate raw. These were the hors d'oeuvres. Our hands were red with seal blood; it even got into the children's hair. Next came the main course, also raw. While the women were butchering the animal and bringing bits of meat, we men were drinking the generous tots of *immiaq* that Augustina kept pouring between mouthfuls. Her husband, a stocky and at first sober-seeming fellow, grew merrier and laughed more loudly after each glass. Drawing his wife close to him, he asked me:

'Isn't she pretty? Plump and dimpled – *mamaq* (tasty)! If you like,' he added, clinking glasses with me, 'you can be *kammak*, the two of you.'

Kammak is a corrupt form of the Danish word *Kammerat* (comrade), here meaning intimate friend.

To my great surprise, Cecilia, who hadn't been consulted, put up such a protest that it was obvious she wasn't joking.

But Jørgensen repeated what he had just said, taking no account of their children, who sat on their three beds and laughed as they listened.

Shy, their mother at first made no reply. Then, urged on by her husband, she told me:

'Yes, we can be *kammak* if you like. And I'll make you fine *kamerussat* for your *kamiks*, since yours aren't thick enough.'*

* *Kamerussat*: outer boots worn on top of the *kamiks* for double thickness and warmth. On the same occasion, I learned that one sealskin is enough to make only one pair of men's *kamiks*.

4

A Greenland Christmas

For days now, the village had been talking about nothing but
Jul – Yule, or Christmas. Apparently it was the country's biggest
festival.

The preparations began on Friday, December 17, when the
villagers scrubbed out their houses using bucketfuls of melted
ice. Though I never saw one of these people having a thorough
wash all winter, they were nevertheless determined to make
their houses spick and span for this great occasion. When
you think of the effort it takes to get water – the number of
trips to the bay to break the ice, the amount of coal needed to
melt it – you begin to have some idea of the importance of
Christmas there.

Augustina even came to clean the walls and ceiling of Cecilia's living room. She stood on a table and wiped the paint with
a rag which she kept dipping first in a bucket of soapy water,
then in some white product out of a can. This gave ceiling and
walls a more or less new look, or at any rate made them shine.

Even at Thue's house they washed and scrubbed the floor.

In another house, Louisa sewed new anoraks for all her family. The thick green cloth was folded double and spread out on
the bed, and one by one Louisa laid out old anoraks belonging
to each person in the house, then cut round them with scissors.
As her husband (another Knud) had his right hand in plaster, she
made the cuff of that sleeve much larger.

'It's only for Christmas,' she confided. 'Afterwards, I'll cut it down to size.'

Every evening after dinner, the pastor taught carols to a group of men and women who met in the schoolroom. A number of young women made handbags, sheaths for pocket mirrors and spectacle cases, all in sealskin, decorated with little diamond shapes of dyed leather, beautifully arranged in mosaic patterns like the ones on women's *kamiks*. An old woman supervised their work.

But the hunting didn't stop for the Christmas season, though it was considerably reduced, either in view of the coming celebrations or because of the ice, which now stretched out to sea and forced the men to carry their kayaks to reach open water. Whatever the reason, we now went hunting only in the mornings; in the afternoon and well into the evening, we played parlor games.

Ice fishing at Ilulissat

At this time of year just a few young men would go for three or four days' camping far from the village to fish for *qaleralik** through holes cut in the ice. On Sunday the 19th, Poyo went off, too, with a team of thirteen huskies and an oil lamp to give him light by night and 'day' during the fishing. As he wasn't due back until the 23rd, and I wanted to observe the Christmas preparations, I decided to go with him on his next trip.

A little after midday on Wednesday, Izâ (Hans and Cecilia's second adoptive son) arrived by dog sled for the festivities. Short and stocky, with thick lips and a friendly face, he walked with a swagger and made the floor ring with his footsteps. Straight away he handed his father a hundred kroner: delighted, Hans showed me the banknote. A quarter of an hour later I met Hans in the village carrying a case of twenty-four bottles of beer. Before I got home he was drunk.

Next day the schoolmaster, who was also scrubbing the floor of his house, told me how a three-month-old baby had been suffocated the previous day by drunken parents lying on top of her all night.

Poyo came back on the evening of the 22nd, a day earlier than expected, with forty halibut. The fish were stored in the larder in front of the house, where our sealskin clothes were also kept, together with a large amount of seal meat, whale meat and blubber. After Hans and Cecilia had gone to bed, we stayed up a long time, talking and drinking *immiaq* . Poyo told me about his fishing trip. At one point he dropped his pants and sat on the privy bucket, still talking away; at another, he squatted on the floor, belching and devouring a slab of halibut, which we ate raw in that stinking atmosphere. You hold the fish by the tail, with the head dangling to the ground over a strip of

---

* Halibut, or *hellefisk* in Danish, a flat fish of the cold northern seas which can grow up to a metre long.

cardboard, and with a knife gently scrape the clinging grey scales. Then you slice into the white flesh. The fish is frozen so stiff that it can't be bent in two. Not a drop of blood appeared when we cut it. The frozen raw flesh of this fish is excellent, but as I ate it I felt I was grinding up sharp little splinters of ice as well as meat. My mouth was numb from it, and I feared for my teeth.

Poyo brought his big black dog into the living room to show me, and we gave him some bread and halibut in our hands. Once we opened his muzzle; a big fang stuck out on either side of the upper jaw. Poyo told me that these two teeth are cut short with the help of pincers, for fear of their growing beyond the lips, making very dangerous weapons if the dogs should turn savage and start attacking men.

THURSDAY, DECEMBER 23. For breakfast we ate raw whale meat, then seal meat with plenty of blubber. Brought straight from the outside larder, the meat was so frozen that, as on the previous evening, it was like chewing splinters of ice.

In the afternoon Cecilia visited the two village stores with Izâ and Poyo to do some last-minute shopping for the festivities. They returned an hour later, laden with pine branches and two cardboard boxes, one containing all the necessary ingredients for brewing forty litres of *immiaq* in time for Christmas Eve, the other holding decorations for the living room. But as the housewives of Rodebay don't start putting up their Christmas decorations until the middle of the night, so as to surprise the others in the morning, Cecilia stowed this box unopened in a corner. From the other one she took out six 65-centilitre bottles of malt extract, a hundred grams of hops, the same amount of brewer's yeast, and eight packets of sugar. Then, while Hans cut slots in a pine trunk to hold the branches, she started preparing the *immiaq*.

She heated a six-litre pot of water, and when it began to

bubble she added the hops. These she let simmer for an hour so as to flavor the liquid. Then she set the pot and its contents aside and placed her big washtub full of chunks of ice on the fire. As the ice melted, Cecilia added further lumps until the tub was three quarters full of boiling water. She poured in the contents of the pot, straining off the hops, then added the malt extract and the eight kilos of sugar. After letting this mixture come to the boil again, she put in the brewer's yeast and skimmed the top of the steaming liquid. Now it was ready. The process had taken five hours.

To improve the quality and make the taste less harsh, Knud the schoolmaster told me that the freshly brewed *immiaq* should be refrigerated to fifteen degrees, then left for four or five days in a cool place, and finally bottled for two or three weeks. But all that seems too elaborate to the Greenlanders, who drink their *immiaq* still hot or tepid – apparently to the detriment of their health.

The baby suffocated by its drunken parents was buried that day, and there was a funeral service at two-thirty in the afternoon. While he was in the school, Knud heard through the open doors of the chapel the sound of the hammer echoing on the coffin lid, as the father of the dead baby nailed it down. On their way to the cemetery, the parents cried their hearts out. There was a big turnout of mourners, wearing everyday clothes.

DECEMBER 24. I woke up suddenly at about four in the morning, and was startled by the sight of the walls of the living room where I slept: the whole place was decorated! Garlands decorated with brightly coloured pictures ran along the walls, festooned from one corner of the ceiling to the other. The pictures on the wall facing me showed young couples holding hands, boys in blue shorts, and girls in full red dresses. On the opposite wall were rows of red-and-white paper hearts. Most of the left-hand wall had been transformed into a miniature Scandinavian landscape showing scenes of rural life: dancers and musicians with faces ruddier than the big red pointed caps on their heads; an old man sitting apart from the dancers brooded on his memories as he smoked his hookah, his face overgrown with a blue beard; next to him were two boys, one feeding bread to two cows, the other giving hay to a horse; facing this group, a cat lapped a saucer of milk; and further away a pig rushed up, while in a doorway a mouse stood on its hind legs in front of a bewildered aproned man standing beside a barrel with beer pouring from its spigot into an overflowing jug on the ground. An apt detail! The wall on the right was plastered with photographs of a previous Christmas. On the floor to the left stood the Christmas tree, decorated with little Danish flags and coloured candles. Another candle, round and red and bigger than a balloon, was set in the middle of the table.

I could hardly believe my eyes. What trouble Cecilia must have gone to, all by herself, to do all that work without making the slightest sound! I was about to go back to sleep when, at about five o'clock and already in her Sunday best, she tiptoed in to put the final touches to her decorations. In the morning we found them dazzling yet delightfully understated – all in all a marvellous display.

8.00–8.45 A.M. Our eyes still heavy with sleep, we attended the first religious service of the day.

10.00 A.M. All over the village the lighted windows shone yellow against the half-light. Smoke billowed from the chimneys, a sign of people melting the gallons of ice water needed for all the coffee visitors would be drinking this day. Stars were still shining. From the window of the school I caught sight of Elias the shopkeeper, beaming as he hoisted a Danish flag in front of his house.

1.00 P.M. The second religious service. All the children from eight to fourteen sat on chairs facing the adults, who sat on benches. The girls were dressed in the national costume, with long white *kamiks* ornamented with lace. The stiff sealskin boots made big pleats at the knees and came right up to the top of their thighs, where the sealskin breeches began. Short legs dangling in the air, the girls tilted their heads and smiled; plump cheeks still glowed from the cold outside, long pigtails were tied with red or white ribbons. Most of the boys wore black trousers, white anoraks and black *kamiks*, but some had swapped the national costume for blue jeans, pullovers and wellington boots. Only a few women, Cecilia among them, wore national costume, each with a handkerchief stuffed into the lining of her right boot at the knee. The men's hair gleamed under its coating of some kind of cream. For the first time since my arrival in Rodebay, Thue crossed the threshold of the church, together with Aqqaluk, Maria and one of his grandchildren. The pastor, arriving last of all, immediately lit up the Christmas tree that stood beside the children. On this day – exceptionally – the men and women sat on the same benches. Part of the service was devoted to questioning the children about their behaviour since the previous Christmas.

5.00–5.45 P.M. The third religious service of the day. All those who attended the first two came along for this one also. The little church was jam-packed and there wasn't a seat left. Seeing me standing up, the pastor got a chair for me. Fresh candles were lit in front of the altar, and where the children had sat in the morning there was now a choir of fifteen women and five men, who rose from time to time to lend a truer note to the carols. After this service, the children went round the houses singing carols and were stuffed with tea and cakes.

Now it was visiting time. Everyone made a point of visiting everyone else. *Immiaq*, Danish beer and *akvavit* flowed. At moments like this, the traveller wishes he had what nearly all Arctic ethnologists have – an isolated house to himself.

*

SATURDAY, DECEMBER 25. At the eight o'clock service, the church was packed again. This was not surprising, for the whole village was now feeling an uncontrollable urge to sing. The hymns and carols rang out more lustily than ever.

Visits began right after church, earlier than the day before.

Hans, Cecilia, Poyo and I ate a late breakfast at the schoolmaster's, then all five of us called on a couple who had taken coffee with us the day before. The husband was a son of the Rodebay policeman. They all – the policeman, his daughter-in-law, his grandchildren, his son and the other brothers – lived together in the same little house. Christmas greetings telegrams from relatives and friends were pinned up in the living room for visitors to read as they entered. And this was not a quirk of the policeman but a strange village custom: people were anxious to prove to everyone that they indeed had relatives, friends and contacts in the outside world. The previous day, in Sophia and Knud's house, I had seen thirteen telegrams stuck to the living-room wall, and in some houses there were more. The host's prestige is reckoned by the number of messages he receives. At the top of each telegram there were three rectangles: one on the left contained the date in red; and one in the middle, the recipient's address; while a third, divided in two by a faint horizontal line, bore the name Jakobshavn above in blue letters. Then came the text, which almost always began with *Asasara*, *Asasakka* or *Asasannguara*, all terms signifying 'Dear . . .' with varying shades of meaning. The message generally ran to several lines, regardless of expense, employing words of staggering length, each one equivalent to a whole sentence. These telegrams were pinned up one above the other so as to display all three rectangles and the whole message.

Late that afternoon the schoolmaster came to ask me to help him lift a man onto his living-room couch who was so drunk he couldn't walk home. I couldn't go right away, as I had all the contents of my suitcase spread on my bed and was trying to sort them out. He went off again, having borrowed my torch, and had only just left when Saqqaq dropped in. Not long after that I took him with me to see Knud, who had managed to get the drunk man onto the couch. The man's face was bathed in sweat, but as we came in he raised his head and looked scornfully at us. It was Johan Dorf, the village headman.

Before telling the unlikely tale that follows, I must remind my readers that Dorf and I were not on speaking terms, because of his refusal to give me shelter when I arrived in Rodebay. In addition, drink turned him particularly nasty.

We left without sitting down. A violent wind, accompanied by driving snow, howled outside until morning.

SUNDAY, DECEMBER 26. Because of the snowstorm, the day was relatively quiet: the exhausted villagers seemed to be taking it easy before resuming festivities. At the church service the *palasi* read the Lord's Prayer in Danish for the first time, but Knud, the only Dane present, was dozing in his chair. His house-boy, Evat, had given him a big lump of seal meat as a Christmas present the day before. As for me, I gave Cecilia a cast-iron stew-pot and Hans a packet of ten cigars.

All Greenland adults, both men and women, love cigars, which they smoke in a very curious fashion, not wasting even the ash . . . After taking a few puffs on a lighted cigar, they put it out by spitting on it, and immediately put the burnt end in their mouths and bite off the glowing ash along with a slice of raw tobacco. They turn it rapidly over and over on the tongue – the cigar end is sometimes still red-hot – and chew it, rolling their eyes with ineffable delight as they swallow the hot ash mixed with shreds of tobacco, then spit out black saliva and announce, '*Mamaq!*' (It's good!) The rest of the cigar is stowed away in a pocket, to be repeatedly lit, smoked, extinguished and consumed all over again. According to them, this warms your face and improves your circulation when you're out in a kayak in the cold.

So Hans seemed very happy with my packet of cigars and, wanting to give me a present in return, that evening brought me a seashell he'd picked up on the shore, and a black ball-point pen, property of the Rodebay school, whose name was engraved on it.

At half-past six I visited Thue, who offered me coffee and a big, hard biscuit. Hendrik and Marianna had moved out with their children and gone to live with Søren Petersen; the house was quieter, but sad and empty. So there was Thue, alone now with his two youngest children and Saqqaq, who had not yet left him. Apart from them, he had his kayak – which never brought back any seal and which he wouldn't be able to use for several months while the sea was frozen over – and his sledge, which had several wooden slats missing. And nobody, nobody in the village offered to help him!

Aqqaluk no longer had a shirt of his own, and had to wear one of his father's. The shirt was far too big for him, so, to make it fit better, the boy, standing in front of me with the shirt on, started to poke new buttonholes alongside the old ones, using a knife. As he cut the cloth, he bent his head down over his chest and kept sniffing in the snot that dripped from his nostrils.

FRIDAY, DECEMBER 31. Starting at one in the afternoon, groups of children made the rounds of the houses, though they didn't sing as they had at Christmas. When they came to us, Cecilia doled out cakes and they left with thanks.

That day produced the strangest sight I'd seen since the start of the festivities. Young men disguised as spirits roamed the village in the endless night. They generally gathered by the roadside in front of the unlit house used as a communal work-shop for building kayaks and sleds. These 'spirits' looked like great bundles of clothes; they were called *mitaartut* (*mitaartoq* in the singular). I was oddly reminded of those grotesque unmasked creatures called *zangbéto*, who in the villages of my native land loom up out of the darkness, creep into huts, and by their unexpected and terrifying presence silence little children who cry in the night. But unlike the *zangbéto* of Togo, the *mitaartut* of Greenland play no useful role in society. They spring out and give you a fright, run after passers-by, perform a ponderous dance, sometimes roll on the ground at your feet, and almost never say a word.

We were at Jorgensen and Augustina's, clasping hands and singing carols round the Christmas tree with the children, when a *mitaartoq* arrived. He was covered in such an array of skins that he must have had trouble breathing; and he had a stick in his hand. As soon as he came in, the terrified children scattered and hid – some under the bed, some behind their father or mother. Hans forgot the pious hymns we'd been singing and broke into a lively, lilting song. The *mitaartoq* answered with his own wild dance, leaping in the air and banging on the floor with his stick. At the end they gave him a cake, and he left saying, '*qujanaq*' (thank you) – the one word he had spoken during the whole performance.

I left Jorgensen's to fetch my tape recorder, so as to record those rousing tunes to which the *mitaartut* dance. But on my way home I was stopped by a man named Eliassen and his wife. He told me, 'You can see the *mitaartut* another time. Come and have coffee with us.'

I followed them back to their house, which turned out to be full of people drinking and enjoying themselves. Their pretty little thirteen-year-old daughter was serving the visitors, helped by her sister, Adina, who was six.

The guests were already half stewed. Eliassen, behaving as if his own wife wasn't there, started to take great liberties with one girl: sitting in front of her, he kept teasing her with shouts of '*Tui!*' as he darted his hand up her skirt. Meanwhile his wife sat down opposite me and gave me a long, slow look; then suddenly her face lit up with a sugary smile, exposing her upper gums and three missing teeth, while blue veins pulsed on her neck and yellowish bosom, where a crucifix gleamed. With the same bedroom eyes, she told me three times, '*Asavakkit*' (I love you), in the hearing of her husband, who burst out laughing at the sight of my embarrassment.

I don't know if *asavakkit* has some more respectable meaning; apart from the girls, Hans and several other men had often said it to me as a sign of friendship. Probably it also means simply 'I like you.'

We all toasted one another with one glass and then another in quick succession, for, besides cups of coffee, we all had before us three glasses containing different alcoholic drinks, and these were filled as soon as we emptied them. In the living room, women enticed me with come-hither stares and suggestive movements.

In the thick of all this uproar, the door was flung open and in came Thue, quite well dressed. How people detested him! They all blamed him for letting me leave his home. They didn't throw him out, but they wouldn't give him anything to drink. He managed a tight-lipped smile and gave each of the guests a long stare. Then, as if asserting his rights, he suddenly grabbed a glass three quarters full of *immiaq* just as we were shouting '*Skål!*' – but he didn't raise the glass to anyone before he started drinking.

Then the *mitaartut* arrived, and among them I recognized Maria; some of them had pulled old pink stockings over their heads, weirdly flattening their noses. But soon we heard the church bell ringing: it was midnight, and all these drunken people set off to the church for the service. No *mitaartoq*, however, took part.

This is when you should see Rodebay – after the New Year's Eve midnight mass. Exhilarated with drink and song, chanting as they roamed through the village, delirious villagers started on another round of visits. I followed the *mitaartut* into about ten houses, then finally stopped at Knud's, where there was still light in the windows. He was with a little man in a white anorak. As he got up to greet me, Knud signalled to me not to speak too loudly: a woman was curled up asleep on the couch in a dim corner of the living room. I took the glass of beer he handed me to celebrate the New Year, and five minutes later set off home to bed.

As I was crossing the yard, I met a man who kept shining his torch right in my eyes. In my country this is considered rude. Who could it be? I turned my own light on him: it was Johan Dorf. Both of us lowered our torches. Thinking no more of it, I went on my way.

I was right by the door of the outside larder when Dorf attacked me from behind and knocked me to the ground. I got up with grazed fingers, and Dorf, like a vicious mongrel that runs off after creeping up and pouncing on a man, scuttled off to Knud's house. I followed him there to have it out with him at once; as soon as I opened the door he swung around, aimed a kick at my belly, and slammed the door in my face. I heard him screaming inside like a madman, 'Get out of here, you rotten nigger!'

It was the first time I'd ever been called that, though I'd long ago realized that when someone having a dispute with a black man calls him 'rotten nigger' or 'filthy nigger' or some such name, it's always some embittered neurotic trying to work off frustrations that have nothing at all to do with the 'nigger'. In this case Johan Dorf, who was envious by nature, was jealous of my success with the women there.

I managed to get the door open again, but he gave me

another kick in the stomach and heaved it shut again. Shouting for Knud, I pulled at the door again, and the bastard gave me a third kick. Trying to dodge it, I slipped and fell on the steps. I was just getting up to open the door yet again when Knud came flying outside.

'Run for it,' he shouted. 'They're fighting inside.'

Winded, and suddenly weak at the knees, feebly I went along.

'Let's go and hide in the school. We're on our own. We're friends, Michel!' Knud panted as we stumbled through the deep snow, floundering into drifts.

We burst into the school building, and Knud double-locked the door. We sat down and he started to cry. Only then did he explain to me that the woman sleeping in his house, so dead drunk that she couldn't walk home, was Johan Dorf's wife.

'But what's that got to do with me?'

All the while, I was trying to control my temper. Knud took my torch to go and see if things had calmed down at his place. He came back very agitated, then went out again. Finally he returned and said we could go back now.

In the yard he had me wait near a shed by the house while he went into the living room. A minute later he came back and called my name.

Two men who had nothing to do with the affair were sitting calmly at the table. The woman in question hadn't stirred: she was still fast asleep. We talked about Dorf's strange behaviour, and I informed Knud and his visitors that I intended to report the attack to the police representative; every village has a man with the imposing title of *Usted bestyrer* – literally, 'village administrator' – whose main function is to keep the police informed. Detectives visiting a community where a crime has been committed go and see him first of all. The one in Rodebay was called Knud Jørgensen.

They told me Knud was falling-down drunk, so seeing that

the whole village was now unprotected by the law, I decided not to be made a fool of: the next time I saw him, I'd teach Johan Dorf a lesson. I was twenty-four. Johan, strongly built and fat as a pig, was nearly thirty-eight. So he wasn't old – he could put up a fight.

Suddenly the door flew open. Dorf came in and made for his wife. I called his name but he just scowled at me, so I grabbed him by the collar. Trying to hold me back, the others hung onto my coat so hard that it ripped, but I took it off and started fighting. The table overturned. Dorf managed to throw himself on top of me, but I pounded his face with my fists, then felt sick at the sight of the bloody mess I made of him. My *kamiks* slipped in the beer; everybody shouted at once. Finally we stopped, and they pulled Dorf to one side. Knud asked him why he had knocked me down and kicked me in the stomach.

'I don't remember,' he said as he slumped gasping onto the couch.

This idiotic answer made me go for him again. I hit him in the face, and we were at it again, till he sat down, winded once more. Suddenly, in came the little man in the white anorak who had been at Knud's an hour or so earlier; I let this little puppet stand there and wave his arms. Saqqaq arrived, learned what was happening, and wanted to give me a hand. Gradually, however, everything calmed down. They showed Dorf the door. He turned back. Knud shouted 'Out!' and gave him a kick in the rear. Some people came to take him and his drunken wife away.

Many of the villagers were afraid of this bully Dorf who, because he ran the Danish shop and handled the trading company's operations in the village (buying seal, fox and other skins), liked to be feared and to play the top dog. But that evening, his fat face puffed and bleeding, he slunk through the village with a lowered head. They congratulated me on teaching him a lesson, and Knud suggested a snack.

A group of children came and sang at the window to wish Knud a happy New Year. He invited them in and handed out cakes. I calmed down at the sight of these children with their peaceful faces.

But that evening still had surprises in store. When the children had left, the five of us sat down round the living-room table and began to pitch into some cold meat. Besides myself and Knud, there were his houseboy Evat, Saqqaq and the young man Pavia. A dispute soon broke out among the three young Greenlanders over the quality of Knud's *immiaq*. Saqqaq found it too weak, therefore second-rate. Knud did not reply, but his houseboy Evat couldn't allow his boss's *immiaq* to be slandered like this, so he and his friend Pavia saw Saqqaq out, but first, to my surprise, they calmly asked him to apologize and to say thanks by shaking hands with everyone, beginning with Knud. And so he did, before he left!

SATURDAY, JANUARY I. We split up at eight in the morning. What blackness outside! When it was time to leave I couldn't find my torch; someone had been seen picking it up after the fight, but nobody remembered who. I went off flanked by Evat and Pavia, who saw me all the way to Hans's door.

'It's for your protection,' they told me.

They were afraid of possible reprisals by Dorf's friends.

Our *Usted bestyrer* never asked me anything about the fight. Recovering from his night on the bottle, he remarked: 'Dorf looks as if he fell out of a *timmisartoq* (helicopter).'

As for Dorf himself, far from making any more trouble, he kept out of my way. And from that day on, every time he had a row with someone, they'd say: 'So you think you can push me around because I'm small and weak? I won't fight – you're stronger than me. But try it with Mikilissuaq and see what happens!'

*

In the evening there was a big New Year's dance at the *Forsam-linghus* (the village hall). I went with Hans, who was not sure where Cecilia had got to. We each took a kroner, the admission price.

The village hall, a small clapboard building, was lit inside by an oil lamp. Hendrik stood at the far end of the packed room and played the accordion: he was the only musician, but everyone joined in the singing.

At first, with pride they sang *Nunarput utoqqarsuanngoravit* (Our Old Country), a fine poem written by Hendrik Lund around 1912, which soon became Greenland's national anthem to a tune by Jonathan Petersen. The poet sings the praises of the *Kalaaleq*, the Greenland man, who survives today and will live on forever, because he can make the most of his country's resources. Then everybody roared the refrain:

> But we must set ourselves new aims,
> And go on growing, that one day we may be
> The respected equals of all other nations.

Hendrik kept squeezing his accordion, but it's not easy dancing to a national anthem! So someone started singing 'Narsaq', a song that salutes the most fertile region of the south, where cows, ponies and sheep graze peacefully in the meadows among clapboard houses. Narsaq abounds with forage and food and green fields lying at the foot of the mountains! The song goes on to compare the delights of Narsaq with those of the biblical lands of Lebanon and Sharon.

During this dance, Søren and his partner Amalia, who had had a few drinks too many, fell flat on the floor. Far from asking if they had hurt themselves, everyone gathered around and held their sides with laughter.

The next song was 'Sunia': When Sunia appears at the far

end of the fjord, towing a whale behind his boat, there is great joy in the village. The Inuit happily divide the animal among themselves, and every villager has as much meat and blubber as he wants! But Niels, Lukas and Markers down so much *mattak* that it sticks in their throats. So they have to be thumped on the back to make all the *mattak* go down! The singers end with the wish that some day Greenlanders will be sailing real whaling ships of their own.

But the liveliest air, started by Hans, was called 'Arnajaq', Snow White! For the Inuit really have a sung version of this fairy tale.* In this song the wicked stepmother, jealous of Snow White's beauty, throws her out in the snow. But the fox, the crow, the snow goose and the Arctic hare lead her to the turf cottage of the Seven Dwarfs. When Snow White gets there she puts the house to rights, scrubs the floor, trims the wick on the seal-oil lamp, and prepares a tasty meal of dried fish, *mattak* and reindeer meat. Then the Seven Dwarfs come back from the coal mine. Delighted to find their cottage spick and span, they adopt Snow White; overjoyed, they begin to dance the rhythmic steps of the *sisamaaq*, the strange dance we ourselves were now dancing. We formed a big circle, and to the music of the accordion took four steps to the left, four to the right, then swung giddily round with our partners. It was a reel, a cheerful fifteenth-century Scots country dance that the natives had learned from European whalers.

Despite all the liveliness and gaiety, I felt a little disappointed. In their amusements, the inhabitants of this western coast have retained hardly anything of their own cultural heritage, nothing that really belongs to them. The accordion which Hendrik tirelessly played was a foreign instrument. As for the Eskimo

* In 1948 the writer Frederik Nielsen published 'Arnajaraq', the Greenland version of 'Snow White and the Seven Dwarfs'. This tale gave birth soon after to a song about the wanderings of the unfortunate princess in the mountains.

drums, made of a circular wooden frame covered with a stretched membrane which is tapped on the edge with a slender stick (strangely enough, never on the membrane itself), nowadays they can be found only at the National Museum in Copenhagen! I rather missed the New Year festivities in my home village, where our dances, not copied from anyone else's, are cadenced to the rhythms of the tom-toms.

To give Hendrik a breather, they brought out records and a small battery-powered record player, and we danced till morning.

On various pretexts, the holiday continued all through the first week of January. On Monday the 3rd, the whole village assembled at the pastor's, where we drank till five in the morning: it was his birthday. The next day was Knud's birthday, and his house never emptied. The following day it was someone else's turn, and so on.

Those who had enough provisions spent only an hour or two a day repairing their outbuildings or kayaks. Only Thue went hunting every day, but only from ten until noon. Today he came back with an eider and a little guillemot. In the dark of their living room, I could dimly see Maria boiling the birds, and Saqqaq squatting down. In the other room the oil lamp was burning feebly, and Thue was sitting on the floor drinking. I had a cup of tea with him while he got drunk on 'Patria', a terrible Danish wine.

JANUARY 5. This evening – unusually – we were all home, and everyone was busy with some kind of work. Cecilia sat on the floor with a thimble on her finger, mending the soles of a pair of *kamiks* with a bodkin, some seal sinew thread and scraps of new skin. She made two patches on each sole, at the heel and toe. Poyo worked on Hans's seal net, while Hans himself concentrated on making a dog whip, which he plaited in a curious way with a single leather thong, so that it resembled a pigtail.

The lash was about seven metres long, tapering at the end. Hans started the braiding about fifteen centimetres from the thicker end, which would later be attached to a wooden handle. In the tough leather thong he made a slit through which he passed the other end of the lash, then pulled on it tightly to make a knot. Then he made the next slit, the next knot, and so on. The knots were very tight and close together. It was hard work, but the result was a series of beautiful interlacing patterns, like little superimposed triangles with the apex of one entering the base of the next, while at the same time a number of elegant wavy lines appeared on either side of the leather thong. To pull the knots as tight as possible, Hans used not only his hands and knees but also his teeth, worn down to the gums: he tugged with them and chewed at the leather, with dizzying jerks of his head. He made sixteen knots in all, and then added to the end of the whip a finer lash of very smooth skin that came from a baby narwhal found in its mother's womb. It is this flexible and hard-wearing section which is used to discipline the huskies. After three days' work, the whip was ready. The thick end was attached to the wooden handle by several layers of cord passing over and under the leather. The whip was nine to ten metres long, and very clean and white now. But wait until it had been used for one winter!

Then Hans made an ice chopper. The handle, two metres long, would later be shortened by twenty centimetres. The blade, flat and rather long, was inserted halfway into the handle, which was split at one end; then it was lashed tight with coiled twine, and reinforced lower down by a nailed iron plate, with six nails on each of the handle's four sides. The bevelled blade was sharpened with a file.

JANUARY 6. There were still a few *mitaartut* to be seen in the village. It all came to an end that evening – it was the last day, the last night, the end of the festivities! Tomorrow the Christmas decorations would be taken down in all the houses . . .

How quiet we all were at home that night! The oil lamp was turned up high, and several candles had been lit at once, as if their combined brightness could drive out our crushing boredom. Hans, who all day long had been alternately sullen and morose, then touchy and surly, slumped in a chair and heaved a loud sigh. He had an Eskimo book in his hand, but he kept nodding off. Cecilia shuttled silently between the living room and the kitchen, where Poyo sat quiet as a mouse. Was it post-Christmas fatigue – here big celebrations are nearly always followed by deep depressions – or was it just the weather? It was piercingly cold now. The Greenland cold, strangely enough, didn't make you shiver or cause your teeth to chatter, for it wasn't just all around you, it was *inside* you. It permeated everything: houses, clothing, people, things. You were reluctant to touch a plate, a pan, a cigarette lighter in your pocket, a watch left at your bedside overnight, and so on.

Yet the Arctic cold is less intense and harsh during the months when the ground is still covered by a thick layer of snow than in March and April when that snow is transformed into ice. Today the temperature, which was minus seventeen degrees in the morning, dropped to minus nineteen. The weather had been mild up to now, considering it was the middle of the Arctic winter. But by January I already felt as if I were living in a refrigerator.

Well now . . . A few days before, old Sophia and her husband Knud had introduced me to one of their daughters, Else, who had recently arrived for the holidays from Jakobshavn, where she worked as a nurse. I spent that night in their house, in bed with Else. The coal stove went out as soon as everyone was in bed, and the inside temperature plummeted. At two in the morning, I still couldn't get to sleep. How cold it was, my God, how cold! A woman's warmth doesn't protect you completely from that sort of cold! Yet we had two blankets and an eiderdown over us, and I was wearing a pullover, too! Else wore nothing!

The next day, January 7, when school started again, the temperature dropped to minus twenty-two degrees. I looked on as poor little children emerged from houses whose faintly glowing windows were curtained with frost, and watched them trudge to school mutely through the frigid whiteness. A great silence reigned in the village. That return to school after the holidays was one of the saddest sights I have ever seen: not one voice, not one sound, not a single dog barking.

The local men, who also seemed to have taken a long Christmas holiday, returned to their hunting that same day. Clad in animal skins, with their ice picks on their shoulders, they headed for the bay, which gave off a greyish haze.

The men had started work again, but not for long, you can be sure. On Thursday, January 13, towards one o'clock in the afternoon, the villagers climbed a mountain behind the village to witness, so they told me, the first appearance of the sun: we did in fact see a faint yellow glow on the horizon, but the sun itself would remain invisible for another few months.

This 'return' of the sun, which seemed to me completely imaginary, gave rise that night to the most curious of the Greenland celebrations, a dance in the town hall that only adult males and their wives may attend – young people are not admitted.

Before going to this affair, Jørgensen and his wife Augustina looked in at our house. Hans, who was getting dressed, appeared in his underclothes and called to his friend's wife. Augustina got up at once and joined him in the bedroom, while her husband stayed with Cecilia in the living room, and never lost his smile. The couple reappeared soon afterwards. This time Hans was dressed and wearing boots – of different colours. Apparently this was done on purpose, because he looked at his feet, clicked his tongue, and gave a sly smile. I couldn't account for this freak of fancy, but Jørgensen and some other men going to the dance had apparently had the same odd notion and were also wearing boots that didn't match.

Hans and Jørgensen took a quick drink and set off with their wives. I followed the two couples as far as the village hall, where a man standing guard at the door refused to admit me. 'You're not married,' he said. But the pastor spotted me from inside and asked that I be let in.

For the first time, I saw coffee being served in the village hall during a dance. The women were no longer separated from the men, as they usually were, but sitting with them. I felt out of place, with everybody paired off from the start. After midnight, with great uproar and loud singing, and faces pouring with sweat, people began to exchange wives. I had read that those present at these public wife-swappings practised what was referred to as 'dowsing the lamps', but that didn't happen on this occasion. You just saw a man get up and go sit beside another couple. He talked for a few moments, then simply left the hall with the other man's wife. Sometimes this happened while the other partner was dancing. The husband thus relieved of his wife looked around him, spotted another woman who might or might not be sitting with her husband, and went up to her. Soon after that, he too went off with a woman who was not his wife. Though the husbands relieved of their mates in this way gave the impression of not being too upset, it looked to me as if most of the women, if you watched closely, were only half willing. Still, like the co-wives in my native land, they seemed resigned to an age-old tradition.

Cecilia went off on Jørgensen's arm! I ran into Hans, who was heading towards a small group at the other end of the hall. Seeing his friend lead his wife away, he told me jokingly: 'You should have got cracking with her, but now she's gone – so hard luck!'

Around one in the morning I left the hall, which was now three quarters empty. At home I was getting ready for bed when the door opened: it was Hans, coming home with Augustina.

She didn't leave until nine o'clock the next morning. Cecilia didn't get home until eleven.

Hans simply asked her: 'Were you cold there in the night?'

'*Naamik!*'

'That's good.'

And that was all! Life resumed or just continued as if nothing had happened between the two of them, who had just openly swapped partners.

So Cecilia was Jørgensen's standby wife, while his wife Augustina was Hans's. Jørgensen was Hans's best friend. Apparently this exchange of wives took place only among friends, obeyed precise rules, and united the two men by an unbreakable tie. Though this practice brought them fresh pleasure, it also involved strict duties, and was not to be compared with the brief exchanges of girlfriends by the young unmarried men. Neither the Jørgensen family nor Hans's would ever go short of food if one of the two men returned empty-handed from hunting, or fell ill, so long as the other had killed a seal.

Motives of survival, then, have given rise to the strange custom of wife-swapping in the Far North. In this light, certain details and behaviour which had made no sense to me a few weeks earlier now appeared in a different perspective. For instance, I remembered Cecilia protesting sharply when Jørgensen told me that his wife Augustina and I should become close friends. A third man – particularly a foreigner – admitted into this union of two couples could endanger, and even destroy, the alliance that would protect one of the two wives in the event (since women don't hunt) that her husband got killed. In the same way, Thue, practically a widower, couldn't enter into the firm and sacred friendship to be formed between two men only by means of their wives, and which laid the basis of un-failing mutual help between the two families.

But if the exchange of wives was a matter only for the couples concerned, why make a public exhibition of it by organizing a dance before the exchange takes place? Probably so as to have no secrets from the village. Hans packed away his boots of different colours, whose significance may be found in the exchange itself, and said to me: 'They fit me like a pair of gloves, but they're sometimes hard to wear!'

For my part, I had grasped the meaning of exchanging wives in the Arctic. All the same, I couldn't help wondering what my father and uncles, my brothers, and above all my grandfather would have to say about it, when I told the tale back home. Perhaps they'd simply think I'd been living among madmen.

'But you're a strange lot, too,' Hans shot back, 'with your eight wives and more under one roof, when it's hard enough living with just one!'

PART IV

A True Greenlander

I

Of Dogs and Men

As Poyo had to return to Jakobshavn, I decided to go with him and pick up my mail. Isolated for the long polar night in a village where, needless to say, there were no newspapers or books, I missed letters so much that it seemed ages since I'd received one. Apart from the many short trips I had made near the village and over the bay to break ice, this trip to Jakobshavn was to be my first long journey by dog sled.

Around ten o'clock we were ready to start. I had put on jeans under my fur trousers, and under my anorak I had two thick pullovers over a cotton shirt. I had given up wearing nylon shirts for good. They chilled my body in spite of all my other thick clothing, and they didn't absorb perspiration. Another thing: synthetic fibres worn with some woollen pullovers have a drawback which I'd come across in Europe but which is much more common in the Arctic: they generate static electricity. I wore two pairs of mitts, one woollen, the other of animal skin (sealskin for the palm, dog fur with long black hair for the back). These outer mitts were a bit too big, and reached down to my knees when I stood with my arms by my sides. Dressed like that, we looked like two big polar bears rearing on their hind legs.

In the entrance, Poyo got out harnesses for eleven dogs. Then he went outside and called them. Mostly they docilely let themselves be harnessed, but some of them snarled, though they didn't run away. Poyo went in among the pack, crouched over each of the beasts in turn and passed the harness over its head, all the time talking gently or scolding harshly, depending on the reaction. Then one at a time he lifted their forelegs so as to settle the shoulders into the harness, an ingenious arrangement of six leather straps. The first strap went loosely round the dog's neck; the second round the trunk; while the third, attached to the first two straps and the others also, was passed along the flanks and round the chest – this last is the strongest strap, the one that pulls the sled (not the one round the neck, which would choke the dog if used for hauling). The fourth strap passed over the back, and the last two under the belly.

When the harnessing was completed, each animal stood at the end of an individual trace which could be up to nine metres long and was attached to the dorsal strap. The eleven traces pass through a chain (sometimes a ring) at the front of the sled. This is the fan harness, the only kind used in Greenland. It lets the dogs fan out on the track, which they can do without encountering obstacles in this treeless land.

Sitting down, ears pricked, our eleven dogs waited for the starting signal. Poyo brought out a soft reindeer skin and tied it on the sledge to cushion any bumps along the way. Then he threw his mitts into the big canvas bag strung onto the back, stretched like a black screen on the framework of the two uprights. This bag contained the provisions and gear for the journey. I put my camera in it, then sat on the sled.

'*Taama!*' (Mush!) shouted Poyo.

The dogs set up a frightful barking chorus, then set off at top speed while my companion ran behind, one hand holding the long whip, the other gripping the horizontal bar fixed a few centimetres from the top of the rear uprights.

Perched on this strange means of transport, you're not quite sure whether to feel sorry for yourself or to save your pity for the dogs trotting in front of you and occasionally turning their heads to look back at you. The passenger is sitting less than twenty centimetres above the ground, and although he is so low, the fear of a spill takes over, at least at the start. The only backrest is the canvas bag, supported by two ropes crossed behind it. The only support for the neck is the bar across the uprights. There is nothing else on this flat vehicle to support the passenger.

Our sled had two main sections: the base, consisting of the two runners, and nine cross-bars. Not a single nail was used in the construction; all the parts were held together by lashings made of sinew, which gave great flexibility when one of the runners passed over a block of ice, a stone, or a lump of frozen earth. Thanks to its upcurved ends, the sled slid easily over these obstacles without injuring the passenger.

After crossing a chaotic barrier of ice blocks all jumbled together along the coast behind the village, we reached the frozen bay. The ice, opaque and even, but cracked here and there, looked like broad slabs of thick glass or caked white mud, all roughly quarried but neatly laid together. After that rough ride over frozen rocks and chunks of ice, it was a joy to be gliding over this flat surface! Poyo jumped onto the sled, where we sat crosswise, I at the back with my long legs sticking out to the left, and he at the front, his legs sometimes to the right, sometimes straddling the runners.

The dogs kept up a steady trot (three others, not harnessed, ran alongside). Their claws made a faint scrabbling on the frozen surface of the bay and sent soft flakes of salty ice flying into our faces. Their gaping jaws breathed steam, and their scarlet tongues, lolling sideways, spattered drops of saliva. Shaggy tails arched back over their hindquarters. Suddenly one of them held back to defecate, but a dog team will not wait. The animal had barely begun to squat before he was pulled forward with a jolt by the others. He gave one yelp and regained his place at once in the pack: one second more and he would have been dragged helplessly along the ice, and sleds have no brakes . . . He had time only to drop a single hard black pellet, and looked as though he hadn't finished. When he rejoined the pack, the dog didn't get his old place back. Certain dogs always like to run alongside a particular companion and won't tolerate an intruder; others have no special preferences in their running mates, but simply dislike being jostled. So the dog's return provoked a short but ruthless clash. The poor beast still kept trying to defecate on the move, tensing his backside in pain. Then suddenly and noisily, from under the plume of his shaggy tail, we saw the eruption of a brownish liquid mess that spattered the ice, but that might well have caught us full in the face if the wind, which was blowing against us, had been a little stronger.

Our weight squeezed a trickle of water out of the ice by friction; it moistened the runners, then vanished in our wake. The whip was not needed for the moment; with its handle looped over one of the uprights, it trailed in our right-hand track and slithered after us over the ice like a long, thin snake. At the point where we should leave the bay, we found ourselves blocked by a channel of open water rimmed with slush. Since the crack was about a metre wide and our sled over two metres long, it was possible to cross over. On top of this fissure a layer of fresh ice had formed, strong enough to support the dogs. They had to cross first, of course, and would haul us out if the ice gave way beneath the sled. A detour would delay us by a good quarter of an hour. Moreover, recent tracks showed that a sledge had passed over this spot a short time before, though it might have been carrying only one man . . .

'Hold tight!' said Poyo, as he whipped the huskies and launched them at the obstacle. The dogs splashed into the freezing water and kept right on pulling the sled. We had just crossed over when the ice cracked behind us, widening the gap by more than three metres.

'It'll freeze over again by evening,' Poyo assured me.

Turning right as we left the bay, we were back again on uneven ground that was bumpy even though covered by a thick layer of snow, its powdery surface blown by the dry wind into hard, crusted ribs. Tufts of yellowed grass and stones brownish with dried moss protruded here and there. The dogs sank up to their hams in the snow; unable to trot over this snowy, stony surface with the same ease as on the solid, level ice, they strained harder, bracing their necks, blinding us with the fine white spray they kicked up in our faces. Sometimes a trace would catch on a stone; then Poyo shouted '*Unigit!*' to bring the team to a halt, and the dogs stopped still and turned eleven snow-flecked muzzles towards us as if to ask, 'What's the matter?' Clinging to the fur beneath their jaws, I could see coatings of snow cemented by saliva and frozen solid on contact with the glacial air. We untangled the trace and set off again.

On the frozen lakes, the smooth snow resembled a white-tiled floor lit softly by the horizon's faint gleam. The dogs pattered rapidly across. With the white of the mountains all around us echoed by white ground under a livid sky, it was like being transported into a world of mist and haze. How could Poyo navigate when everything around us was so blurred? I didn't know, but for now some old tracks showed the way, and we simply followed them.

From time to time, Poyo jumped down beside the sled, ran at top speed beside it, then bounded on again. I couldn't understand why he should push himself so hard, instead of just sitting quietly beside me. It was very cold. Already my feet were numb in my *kamiks*, and I grumbled about it to my companion.

'You're cold because you sit still,' he told me.

And so I learned that during a long journey by sled, a passenger should not remain seated all the time as if in a car. Like the sled driver, he has to make more than a quarter of the trip on foot, running to keep himself warm. We took turns to run beside the sled, and whenever I jumped back on after a gallop through the snow or over the ice, I felt an intense warmth flood my whole body; if I opened the neck of my anorak, a gust of warm air puffed out. I realized then that supposedly warm clothes don't really give us any heat: they simply help conserve the warmth from our bodies. Yet, paradoxical as it may sound, the more pullovers you wear under your anorak, the more you sometimes feel the cold. There has to be a space for warm air in between your body and your clothes.

Once, as we were crossing a frozen lake, we both leapt onto the sled at once after running alongside for several minutes. Hearts hammering, we lay on our backs, gazing up at the sky. After a moment, Poyo stopped talking and closed his eyes; he seemed to be dreaming. I too let myself drift off into this pleasant idleness. We were surrounded by deep silence, broken only by the faint, even hissing of the runners. However hard I listened, I could no longer hear the scrape of the huskies' claws on the ice. Whether or not they still kept looking around to see what was happening on the sled, they remained linked as if by a sixth sense with the driver, respecting even his rest. So for a quarter of an hour they drew us along like this, without the slightest jolt, or any fighting or barking. 'Keep quiet now, the masters are sleeping,' they seemed to have decided. If during this dreamlike period they had come across a polar bear, a channel in the ice, or any other kind of danger, they would soon have sounded a warning. Who said huskies aren't intelligent?

A light snow began to fall. We roused ourselves and sat up. Poyo really had been sound asleep!

After a little more than an hour's travelling, we halted to unravel the traces that the dogs had tangled by changing places too often. While Poyo sorted them out, I reached for my notebook to make some notes and found, to my surprise, that not one of the four different colours in my ballpoint pen would write: the ink was frozen! My handkerchief was like a sheet of crumpled tin. My folding camera wouldn't work. I should have kept these objects under my clothes, against my body, to prevent them from seizing up.

A little later, as we were toiling up a steep hill, we suddenly saw another sled speeding down from the top. Poyo shouted to his huskies to keep to the right (*'Ili-ili!'*), the other driver shouted the same to his dogs, and the two sleds passed on each other's left in a flurry of powdered snow. We struggled to the top of the hill, and then it was our turn to start the dizzy plunge down the other side. Trying to cut our speed, Poyo clung to the uprights and dug his heels into the snow for brakes, but the effect was merely to whip up clouds of snow. The sled went bounding down the hill, sometimes drawing so close to the dogs that I feared it would run them down. No longer heeding orders, they seemed instead to be driven frantic by Poyo's shouts to slow down, and charged on downwards. Swept along in this mad rush, I felt sure we would break our necks at the foot of the precipice. Just then, in fact, the sled suddenly skidded on a stone halfway down. I lost my balance, was flung off, and rolled head over heels to the bottom.

Still gripping the two uprights, my companion brought the sled to a graceful halt beside me, while I wobbled to my feet and dusted the snow from my clothes. He didn't even ask if I was hurt.

'Tupinnaq!' (Unbelievable!) he exclaimed. 'How can a man fall off a sled? It's not possible, yet you, you managed to do it. I saw you rolling down like a seal's bladder, and I couldn't believe my eyes!'

As he said this, he held his sides and laughed until he cried. If I'd broken a leg, he'd have had much the same reaction. That's Greenland courtesy: no kind words or consideration for anyone who falls down. I cannot remember a single instance when the sight of an accidental tumble – whatever the outcome – didn't cause howls of laughter. Therefore, dear reader, if you happen to go to Greenland, be careful not to stumble. Your first aid will be a burst of general hilarity – even if you break a rib.

We got onto the damned sled again. Our path, which consisted only of previous sled and dog tracks, was a little over one metre wide. Sometimes it disappeared behind a hill, then appeared again, visible over a great distance as a narrow grey ribbon winding through that vast white space. It would take only a snow squall to obliterate it, and we'd lose our way in this featureless land. And although our field of vision was often very extensive, I was much more afraid of getting lost here than I would have been in a forest.

The track was now crossing a landscape of hills and frozen stream beds. Whenever Poyo felt any doubt about the state of the terrain beyond a rise, he ran ahead and climbed up some hillock to reconnoiter. Every now and then the dogs started whimpering and refused to pull the sled, sensing our lack of assurance. On the steep slopes of the deep ravines which we descended on foot to obviate a fall, my companion would shout 'Qaagit!' to the huskies, who stopped and turned the sled round. He then started cautiously down the precipice, leaning back against the uprights of the reversed sled, while the shrewd dogs inched down after him. When we had to scale a rise, the dog team took the lead again. We would climb the slope clinging to the uprights of the sled, which hauled us up.

In one big ravine we passed a hunter's camp – a sure sign that we were nearing Jakobshavn. The camp consisted of a single wooden shack, with long icicles hanging from the roof. The door and window were crusted with snow. Obviously, for the moment there was no one in the isolated cabin.

A little later we were moving slowly across the open country and frozen marshes outside Jakobshavn. Arriving in a town or village is possibly the high point of the driver's journey, so it has to be spectacular – the sled should drive at top speed through the streets, and above all the dogs shouldn't show the slightest trace of fatigue, so that people will say: 'He's got the best team in the region!' So Poyo stopped in the plain about a quarter of an hour's drive from Jakobshavn to give his dogs a breather. Although Greenlanders love to make fun of others, they fear ridicule themselves. He inspected the traces and the general condition of the sled, then took his stubby pipe out of the canvas bag and contentedly began to smoke some cheap tobacco that stank.

After that halt, we did not stop again. Sometimes the whip hit a dog's ear and the dog let out a yelp; then all the others would lower their heads for fear of the same treatment and put on a burst of speed.

We raced round a bend, raising a cloud of snowdust behind Knud Rasmussen's childhood home, then swept down the main street in a chorus of barking and whip cracks. Townspeople turned round to nod their approval. But at the second bend, to the left, the far right-hand dog fanned out a bit too far and caught his trace on a telegraph pole. The trace snapped with such a shock that the dog went flying about six metres from the sled and rolled over and over in the snow. Swiftly it jumped up again and scurried whimpering after us. Unfortunately for Poyo's ego – and for the husky's – there were bursts of laughter as we passed by.

The sled drew up behind Cecilia's sister Nikolina's house, and children swarmed out to greet us. Having left Rodebay at ten, we arrived here at one-thirty in the afternoon: we had gone twenty-three kilometres as the crow flies in three and a half hours. Not all that far, but considering the rugged path we had to follow, the intense cold, and the thick snow on the track, I couldn't help admiring the dogs, without which a man would be more helpless here than a city dweller without a car. We left them harnessed outside, where they sprawled in the snow while we went into the nice warm house.

Straight away, Nikolina served us a pile of boiled seal meat, which we made short work of. Gideon, Nikolina's husband, then arrived; without a word, he headed for the other end of the room.

'Say *gudao*,' (good day) his wife prompted.

Nikolina's intervention was due entirely to my presence. Usually, when a Greenlander arrives as we had from another village, he enters the house, sits on a chair or on the floor, eats at once if he feels hungry, and exchanges only a few words with his host about the state of the track, the dogs, or the weather – all that without even a hello. In my own land it's exactly the opposite: for example, an uncle, aunt, or anybody else who enters my father's house must first shout a greeting while crossing the yard. The entire household choruses a reply. The guest is given a seat under the awning, somebody brings water, and he drinks and rests for a while. Then, the wives and children file past to greet him, and he replies. After a few moments, he gets up and goes to greet the wives one by one. Only then is he brought into our father's living room.

Gideon gave us a curt '*gudao*,' which anyhow is a Danish term, a popular corruption of *goddag*, good day. Then Poyo and he began to talk.

'The track is in good shape . . .' my companion began. 'But there's a hole in the ice at such-and-such a place . . . It was fairly windy . . . It snowed . . .' And as he talked, he chose the exact word to describe the particular hole we saw, the particular kind of wind that blew, the kind of snow that fell. For me, there was only one kind of snow; not so for an Eskimo.

We stopped in at the post office, where there were five letters waiting. Then, after buying eight bottles of beer for the return journey, we left Jakobshavn in the deep dark of half-past three in the afternoon. Ahead of us hung the crescent moon. When we reached the plain, we each drank one bottle of beer, promising ourselves not to have another till we had passed the deserted little shack. When we opened the bottles there, I was astonished to find that the two bottles, which had taken a good shaking from the jolting sled, did not foam over. When I casually tilted my head to drink – surprise again – nothing came out of the bottle! The beer had frozen into a slush, which we slurped up vigorously.

My right foot, which hung over the edge of the sled, banged hard against a stone, but it was so numb with cold that I felt no pain at the time. Even when I began to warm myself by galloping through snow halfway up to my thighs, my foot still didn't hurt.

Not until we had completed a moonlit passage through a particularly dangerous ravine did we open another bottle of beer. This time it was frozen so hard that Poyo smashed his bottle to get the solid drink out, and I did likewise. We wiped the beer and scraped it with a knife to get rid of glass splinters, then sucked it like a lollypop.

We were having a lot of fun, but not for long. About ten kilometres from Rodebay, clouds came scudding across the moon, driven by a furious wind that began to lash at the ground around us, howling and scouring up the snow. We shielded our faces with our hands to avoid being blinded, but soon the wind reached blizzard strength. No more moon. No more anything. Our torches only accentuated the darkness by contrast. Poyo relied on his native instincts and kept going for another ten minutes, urging on the dogs, but he soon realized we were off the track, which the snow had obliterated. Aware that we were lost halfway to Rodebay, without a trace of a landmark in the blizzard and the dark, I could not suppress the vision of an icy death. On my own, I would have panicked, but Poyo kept his head and did the only thing possible: set about making camp. With his knife and bare hands, he started to dig in the snow but soon realized that we were on a lake. He managed to get his bearings, and we headed for the nearest shore, where we discovered a hill, and worked our way around to its sheltered side to get out of the icy wind. We dug a hole in the deep snow and huddled inside, and Poyo made the dogs lie on top of us for extra warmth.

The squall lasted all night, and we didn't reach Rodebay until the next morning at about eight o'clock. Back home, when they pulled off my *kamiks* for me, my feet seemed dead. I sat by the stove to thaw the ice which had formed on my moustache. A little later, in the warmth of the living room, I began to feel unbearable pain all over my feet, but especially in the right toe, which was all swollen: it was this one which had struck a stone the night before. I had just received my baptism of fire – or, rather, of frost!

In the days that followed, I had great difficulty walking on my swollen feet, and I was to become the laughing stock of the village. It cut me permanently down to size with my 'enemy' Johan Dorf, especially when he learned that I had fallen off the sled!

On the morning of our return, Poyo attended to the dogs, which hadn't eaten for twenty-four hours. He brought out a barrel full of halibut which were frozen so hard and tight that he was forced to chop the barrel in two with a hammer and chisel, then put the frozen block beside the stove to soften the contents, before throwing them to the dogs. As I watched them pounce hungrily on the fish, I felt a surge of sympathy for the beasts who had saved our lives – or at least mine.

Yet various other experiences were to reveal more of their

complex character. For example, once the schoolmaster went to break some ice in the bay. While he was tackling an enormous block, his dogs, which he hadn't thought to tie up, ran away with the sled. All Knud's shouting and gesticulating were a waste of energy. The beasts were chasing another team, which fortunately was making for our village. The dogs came panting up to the schoolmaster's house and lay down calmly in the snow in front of their pen, apparently not giving a damn about their master. Raging mad, Knud had to come back on foot. He drove them straight back to the bay and gave them all a good whipping at the scene of the crime – in order, he said, to make them understand why he was flogging them.

This incident happened only a kilometre from the village. If it had happened further away, in the vast frozen wastes, the dogs would have come home just the same, leaving their master to die.

Knud told me: 'Though they're stronger than European dogs and can stand cold and hunger longer, they display much less intelligence and have only three things on their mind: feeding, fighting and fucking. Just last night they ate a whip I'd left outside, then had a savage fight with another pack that came looking for their bitches. A lot of blood was shed. By the way, I must show you something . . .'

He went into the bedroom and came out carrying a dogskin with a thick pelt of beautiful red fur tipped with black. But it was terribly gnawed around the edges and badly damaged. It was the skin of one of his own dogs which had fallen prey to some other huskies the year before; they had ripped out his chest and eaten it along with the head and forepaws.

But it was not until two months later that I learned the full ferocity of huskies. It was in Jakobshavn, where I was staying with Gideon and Nikolina. One day I saw a funeral procession passing by – fifteen to twenty people walking towards the cemetery with a white coffin carried on the men's shoulders. They were burying the remains of a man of fifty-five, Hans Gundel, who had been torn to pieces by dogs. Had he been drunk, as the locals claimed, or had he suffered a heart attack, as was officially reported? Whatever the reason, the dogs had made a meal of him. It had happened on a Friday night. The dogs had gone for his throat, then devoured his stomach, face and legs. Strangely enough, they hadn't touched his *kamiks*. They found the feet nearby, and it was his boots which enabled the police to identify the skeleton. The clothing had been torn to shreds to get at the flesh. One arm was missing. Ten dogs were slaughtered that Saturday, two of them belonging to the victim. As is the custom, since the country's freezing climate delays decomposition, Hans Gundel's remains were kept for three or four days by his family before being buried. It was the only such accident in Jakobshavn that winter.

2

The Boy Who Killed a Fly

The frequent sudden changes of temperature in winter didn't bring just unpleasant surprises, and it's worth recording one special occurrence, rare in the Arctic.

Halfway through February, when the sun had not yet returned to the Arctic and Europe was still in the grip of winter, we had a few days of relative warmth in Jakobshavn, when the weather was suddenly mellowed by a warm, blustery wind that caused a long break in the ice. True, the temperature didn't rise above zero, but for people who had been suffering the full rigours of Arctic cold, this slight variation was enough to produce an impression of gentle warmth. On February 15 you could see the inhabitants dressed only in trousers and just a cotton anorak, without pullovers.

As an Arctic greenhorn, I remarked to Gideon: 'Perhaps the ice is going to break up, and we'll have an early summer.'

'Oh no!' he replied. 'Just wait a bit and you'll see . . .'

On February 17, the weather was still glorious: I could even go out in my shirt sleeves! Better yet, the opening in the ice grew wider, and icebergs were seen floating down it towards the island of Disko, invisibly propelled by this warm wind, which continued to break up the ice-pack and to extend the first fracture by creating others running in all directions, till by February 19 the bay was almost entirely free of ice. Only a narrow strip of dirty ice still clung to the coast, while the free waters of the open sea, rippling and dark blue, sparkled all the way to the island. But there is nothing predictable about the Arctic.

The fine weather lasted ten days, then on February 25 it stopped as abruptly as it had begun. The biting cold and glacial wind soon forced us back into those bulky animal-skin garments which I had mentally cast off for a few months. By the 26th, the sea was completely iced over again. It had happened so fast that I couldn't believe my eyes. After all, how could that expanse of water, still rough and ice-free only the day before, have become completely frozen overnight? Not until much later did I learn that, through the phenomenon of supercooling, 'a body can on occasion remain liquid at a temperature below its freezing point.' In other words, the rapidity of the freeze can be explained by the fact that in the Arctic the sea, when it thaws in winter, still contains a large number of ice crystals which are quickly transformed into compact ice by any further fall in temperature. In the end, all you can say is that such ice is gone today but here tomorrow.

In less than twenty-four hours, the returning ice was thick enough to halt eight fishing vessels as they headed for the coast. The most powerful of them just managed to break a channel, and the others followed slowly into the bay, where they anchored one behind the other. That same night the channel froze, closed in on their hulls, and held them prisoner.

The temperature dropped back to minus forty degrees, so that in mid-March – while winter was drawing to its end in Europe – the Arctic faced intense cold once again. It was then that I remembered Gideon's reply. The sea was once more a vast, white, solid expanse in which captured icebergs alternated with snow dunes polished by the wind and ice hummocks that looked like frozen waves.

The canted ships caught in the pack-ice were like wrecks half buried in the sands of some immense white desert. There being no playgrounds or open fields in the neighbourhood, to my astonishment the young men began playing soccer on the frozen bay during the few hours of light we had each day.

In theory, the sun does not return to this part of the Arctic until late April or early May. Except for the diffused light that reached us from below the horizon between one and three in the afternoon – without our ever glimpsing the sun – we were still deep in polar night in the month of March! Every day, during the two hours of half-light, little groups of fur-clad men, like ghostly shadows among the icebergs across the bay, could be seen fishing through holes in the ice, their dogs, still harnessed, crouching beside them.

One day, intending to take some fine photographs of fishing through the ice, I walked over to these men with my camera. After taking the first shot I could no longer move my fingers: they hurt me as if they'd been burned or cruelly squeezed in a vice. I quickly put my mitts back on, then slowly and painfully flapped my hands to get the circulation going again. This amused the Greenlanders, but I couldn't even join in the laughter because the bitter cold made my lips crack with the first hint of a smile. However, as I could only operate my camera bare-handed, I took off my mitts again and managed, after great persistence, to take three pictures in a cold that defied description.

And yet, it was in that frozen hell that these cheerful men watched patiently for food beneath the ice! Eventually I began following my hosts to their work. Every day towards noon, just before the day-gleam lit a corner of the sky, the three of us – myself, Gideon and his second son Justus, aged nineteen and blind in one eye – made our way onto the pack-ice, out among the icebergs. Posting ourselves beside a narrow channel, we fished with a line, catching some *uuak*, a kind of bull-head with a big head and a wide mouth, and some *kanioq*, a spiny fish, all bones, and smelly when cooked (you have to be really hard up to eat it!). How can fish exist in that freezing water?

Each time we caught one, we laid it still wriggling on the ice and chopped off the tail with the ice chisel to drain its blood. I was amazed to find that Greenlanders, who can drink seal broth black with blood, dislike seeing even a single drop of blood in a fish. When I expressed my surprise to Gideon, he admitted that he didn't know the significance of this practice, though he had always observed it. Perhaps the custom can be explained by the fact that Greenlanders were originally meat-eaters and didn't like fish. Since for most peoples blood is the soul of all living things, a fish drained of its vital fluids was no longer considered such but became just another kind of flesh. More or less analogous substitutions can be found in our African societies, where in eating an animal prohibited by tradition we mentally replace that animal by another. Or else, in ritually consuming a forbidden wild beast, we do not believe that we are really eating the animal in question: instead, we are absorbing strength, valor, cunning, in other words, the attributes that once were his. Admittedly, the comparison with draining fish of blood is a little far-fetched, but substitution is common to both cases.

On days when the fishing was good, we ate a few fish raw, right there on the ice. You may wonder if raw fish tastes good. Yes, it does! And I would gladly eat it again, but only in the Arctic of course! Besides providing the raw fat which the body needs in order to help resist the cold, raw fish exposed to glacial air is firm, even hard, and doesn't smell. It is wholesome and pleasant to eat, even when crunchy with ice crystals. However, I would never eat raw fish in my own country, for in our hot climate it goes soft and limp and starts to smell within two hours' exposure to the air.

As for seal blubber, that native delicacy, it is simply nauseating for a foreigner and resembles tallow. Lightly dried and yellowed by the sun, then 'hung' as the Greenlanders like it, it smells rancid. And when frozen, frankly it even tastes like candle wax.

One day there were about twenty of us fishing on both sides of a channel. Gideon was catching *uuak* after *uuak*, while the fisherman opposite him caught nothing. Justus took this man's place and began catching *uuak* just like his father. But what struck me as strange, to say the least, was that the spot where Justus had settled was considered for the time being as the 'property' of the unlucky fisherman who had been there before. So Justus found himself having to give him the first fish he caught. Then, when he went on catching them, he had to throw one in his father's bucket, the next in the fisherman's, and so on – while the other fisherman, now fishing just a little way off, enjoyed his same bad luck. At the end of the afternoon we went home with our plastic bucket full of fish. But on some days – as if bad luck had overtaken all of us – the catch was so poor that the three of us brought home only two or three sorry *kanioq* to feed the whole family.

Sometimes Poyo sledded in from Rodebay so that we could go fishing for halibut in the great frozen fjord of Jakobshavn. He used to arrive late at night, when we were all already in bed. Softly opening the door, he crept in without waking anyone and, still wearing his furs, curled up on the floor at our feet, on a corner of the reindeer skins we slept on.

The next morning, when we woke up at six, Justus and Poyo would head straight for the kitchen without bothering to wash; picking up a knife, they would squat down to a hearty breakfast of slices of very cold raw meat that gave off a musty smell. On other occasions, it might be the lungs of a seal or of a white whale, also raw and frozen. I never felt hungry enough in the morning to share this copious meal but made do with a cup of coffee. My companions had coffee too, with slices of bread and butter. Then about eight o'clock we set off in sleds, sometimes with other young men of the village, for a camp some way from the frozen fjord, where we fished all 'day' by the light of our oil lamps, sleeping at night in a cold little hut or in a tent set up on the frozen sea itself! After two or three days we returned to Jakobshavn, sitting on piles of frozen halibut heaped up on the sleds and covered with a reindeer skin.

Poyo sold three quarters of his catch, sometimes all of it, outside the Danish store in Jakobshavn, then headed back to Rodebay by sled.

One afternoon, as he was about to set off for his village, a girl came and asked if he would take her to Rodebay. Poyo gladly agreed. Watching them set off, Justus grinned: 'They're going to make love on the way.'

'Impossible!' I replied, having in mind not only the bone-chilling cold but also the thickness of the clothing we wore, and especially how hard it was so much as to piss when you travelled by dog sled, since fur trousers have no centre fly, only two openings at the sides.

'Oh, it's easy, love in a sled!' they all chorused – Justus, Gideon, Nikolina and the girl's father, who was with us.

Once or twice a week, when we had enough fish for ourselves, we fished in the bay for *eqalussuaq* – the blue shark, used for dog food. To do this, we kept clear of the open channels, digging a hole in the ice about half a metre wide, and dangling a long, thick line equipped with secondary lines right to the bottom of the sea. These smaller lines carried hooks baited with scraps of meat.

After unwinding the right amount of main line into the water, we tied it to a long wooden pole placed across the hole. Then we waited, keeping the hole open by breaking the thin ice that kept forming on the surface. Sometimes we waited for two hours like that, sitting on a block of snow or ice.

A sudden jerk on the line alerted us, and we started to haul in. As soon as the long, grey body appeared, one of us would shout: 'We've caught an *eqalussuaq*!' At once the others left their holes and ran over to lend a hand, though mainly they wanted to see the catch with their own eyes and offer endless advice and comments.

We hauled the struggling shark out onto the ice. His spectacular jaws, forming a half-circle below the snout, almost in the throat, opened and shut with a loud clash of powerful teeth. We slashed him open with a knife, giving him part of his own entrails to chew on – to keep him busy, my companions said, while we were finishing him off, and so to prevent him from trying to bite any of us, the blue shark being reputedly the fiercest of all the shark family. Once he was dead, we cut off pieces of his steaming flesh, which is considered the best bait for catching other sharks. Beside the hole, the dogs lapped greedily at the blood on the ice, then polished off the head and part of the entrails which we threw to them. The rest of the creature was cut up on the spot into long strips which would be dried before being given to the dogs.

In communities situated on Disko Bay and further north, these different kinds of fishing through the ice sometimes went on until April.

Since the previous June, I had covered the whole south of Greenland, then Disko Bay, which can be considered the borderland between the South and the Far North, and I had come to know the people and customs of the various localities. Now I felt a longing to go to Thule, the most northerly 'town' in Greenland and the whole world, and spend the rest of my life among the real Eskimos – an inexplicable, muddled, yet vivid wish, triggered perhaps by the pursuit of a recurrent dream, the lure of the distant unknown, or quite simply by the fatigue caused by constant travel, and a desire to find some last fixed point which would be neither southern Greenland nor Africa, and above all not Europe! But because the north was still inaccessible by sea because of the pack-ice, even though my mind was made up, I was stuck in Jakobshavn for the time being. A sled journey to Thule – more than a thousand kilometres away – would be not only expensive but dangerous. So I waited.

On March 14, the reopening of old channels finally allowed the eight fishing boats to break free of the ice. That same day the Company boat made its first trip of the year to Rodebay, returning in the afternoon. Two days later it was ready for its second trip, to Claushavn (Ilimanaq), a village of two hundred and fifty inhabitants, twenty-seven kilometres south of Jakobshavn. Sometimes it was necessary to go south in order to make travel connections for the north. So I took the Company boat and left Jakobshavn on the afternoon of March 16.

At Claushavn the pilot, Evat, radioed the Company office in Jakobshavn for permission to continue as far as Christianshåb, thirty kilometres further south, as the boat's cargo included barrels of oil meant for that village.

'The ice is quite broken here,' Evat explained, imagining that it would be the same throughout the area. There were only slabs of ice separated by numerous channels, and they had offered the ship no serious resistance. So the company gave permission.

'Come with me to Christianshåb,' he said. 'It's a big town, twice as big as Claushavn; you'll have a better chance of finding a ship for Thule there.'

In that part of the country, weather conditions were sometimes purely local: they could vary considerably from one village to the next and defy all forecasting.

So, only a quarter of an hour after anchoring at Claushavn, we set off again, intending to reach Christianshåb by evening and spend the night there. The sea was calm and fluid and right into the distance the ice was nothing but a thin, greenish film, transparent and flexible, that followed the movements and contours of the swell. It was like an immense veil laid on the water, so flimsy that even at its thickest it broke beneath our bow wave before the bow could touch it.

We had no trouble at all for three quarters of the trip. But an hour out of Christianshåb, the ship suddenly ran into the toughest ice-barrier I had ever seen. How could pack-ice which was broken up to the north of Christianshåb, and in places non-existent, be so thick and solid only about sixty kilometres further south? After some futile attempts to break through, the ship stopped: we were reduced to spending the night in the freezing cold and only eight kilometres from Christianshåb, whose lights we saw glimmering just ahead. Getting down onto the ice, we planted planks of timber in it to act as bollards for mooring the ship.

At seven the next morning, after a dreadful, sleepless night spent in the chilly little cabin under the poop – it was heated by a small portable stove, and we'd have been warmer inside a refrigerator – we unloaded the oil barrels and other goods onto the ice. Sleds arrived shortly afterwards to shuttle them to the village. Rather than earn himself some money by moving a barrel of oil, one of the young sled drivers decided to carry my luggage instead. Josef (as this friendly young twenty-year-old was called) was pleasantly surprised to hear me speak his own language and gave me a warm welcome. He loaded my rucksack and suitcase onto his sled. I sat beside him and we set off unhurriedly for the village, chatting so loudly and cheerfully that we might have been old friends. Then I asked Josef if he could put me up.

'Of course!' he answered, with a joyful smile that I am never likely to forget.

We were still on the bay when I caught sight of the village, which was clearly divided into two distinct parts: all the houses on the left looked new and were built on a plateau; all those on the right were old, even dilapidated, flat-roofed and mostly painted red, backing onto a high chain of grey mountains that snaked away like a high wall. A broad stretch of flat wasteland, where rainwater must stagnate in summer, separated these two parts of the village.

With a supreme effort by the dogs, the sled left the bay and climbed a steep slope leading to the main street, smooth with ice. We followed the street to the edge of the plateau, where we stopped outside a house by the road. Josef, whose father was dead, lived with his younger brothers and sisters and his mother, Martha.

When he entered the house with my suitcase (I had stayed outside to unload my rucksack), his mother was intrigued.

'You're not alone?' she asked.

'*Naamik.*' (No.)

'*Kinalu?*' (Who came with you?)

Josef was dying to tell the news to his family, but he didn't answer straight away. Like a good Greenlander, he kept his excitement under control. I heard his mother repeat her question:

'Eh? *Kinalu?*'

Josef finally lets fall the word:

'*Qallunaarlu!*'

There was a trample of feet on the floor, the door opened, and the whole family – even little two-year-old Søren, Josef's half brother, the son of a Danish workman – came rushing outside. They stared at me in silence. Once they were over their surprise, Jonas, Rasmus and their sister carried my rucksack inside among the three of them. Up to now Martha, their mother, had said nothing. But as soon as her children accepted me, she was bound to comply with their wishes and take me in. As I have said, words can't describe the total freedom children enjoy in this country. It is they, rather than the adults, who first adopt the stranger. So the great merit of this people is that in all cases where intuition counts for more than reason, they always recognize and follow the natural instincts of their children.

When my adoptive father heard a few weeks later about Josef's cordial welcome, he wrote asking what my Greenland friend would like as a souvenir from Paris.

'A postcard of the Eiffel Tower,' Josef replied without hesitation. 'I'll put it up on the living-room wall for our visitors to look at.'

On May 7, we received a large envelope containing not one but ten large postcards of the most famous Parisian monument. Josef was in seventh heaven.

Poor Josef! He was not fated to decorate the living-room with the marvellous cards as he had planned, for the next day someone came running to the house to tell us that Josef had just been rushed to hospital. While painting the roof of the local factory he had fallen from a height of about ten metres. Horrified by the news, I ran all the way to the hospital. As I arrived I met the doctor, who told me: 'Our small medical post can only give first aid. Josef will be moved to Godthåb or Egedesminde at the earliest opportunity. I have asked for a helicopter.'

Josef left that afternoon, and I didn't see my good friend again for some time. It was only an ankle that he had broken.

On May 1 the whole village celebrated May Day, the Festival of Labour, which like all the festivities there started indoors.

Then all at once, at one-thirty in the afternoon, demonstrators took to the streets with placards reading, 'Equal Work, Equal Pay!'

The Greenland workers were demanding the same wages as Danish workers. Altogether there were about two hundred people marching, nearly half the population. The procession walked slowly behind a green truck which moved at a funereal pace, displaying a big Danish flag on each side. Inside the vehicle were four accordionists and a guitarist. In spite of their demands, the people who made up this cheerful march, which actually included more children than adults, were so jolly and peaceable that the police felt no need to intervene.

When they drew their pay the next day, Danish and Greenland workers alike indulged in their favourite pastime – getting drunk. But during the night a young Greenlander was shot dead by a young Dane who was engaged to a beautiful girl from Ammassalik (Tasiillaq), on the eastern coast; the girl was said to be the cause of the trouble. Hearing of the murder of their fellow countryman, the Greenlanders didn't turn against the Dane. There wasn't even a protest. But isn't silence sometimes a form of protest? The consensus was that the murder was caused by drink, but opinion was divided in Greenland households: some natives, mostly the younger generation, believed that the Dane had killed in cold blood – yet they didn't cry for revenge. The rest – the adults and especially the old people – didn't lay the blame on the young Dane's excessive drinking or even on jealousy but, strangely enough, on 'the return of the sun, which by its sudden and brilliant appearance after months of darkness can create in foreigners a state of euphoria and agitation – in other words, loss of control'. So the young Dane was not responsible for his actions, according to these old folks, who even went on to say that he should be released, to give him a chance to do good deeds in future and so make amends to the community that he had robbed of one of its members. This explanation of the influence of the sun on the young Dane's actions may appear naive, but considering the fury such a murder would have unleashed in other lands, particularly in Africa, one couldn't help appreciating the wisdom of those old Greenlanders, who, with their imaginative interpretation of a natural phenomenon, appealed to the young for tolerance and non-violence. What a lesson!

The young Dane was held in the one cell of the tiny police station; a few days later he was repatriated under escort to Denmark to stand trial.

It was seven o'clock on the morning of March 17 – while we were blockaded by the ice off Christianshåb – when I saw the sun over the horizon for the first time that year. Its rays shone just as red and brilliant as in the magnificent sunrises of the desert, flooding around us and glittering off the vast expanse of ice, which exhaled a rainbow haze.

After that, the sun stayed in the sky a little longer each day. Now there was daylight, and if not continual daylight, at least there was no longer the polar darkness. The sun still set, but the sky where it disappeared was purple and full of promise for the morrow. Each day brought an imperceptible change, and the sweep of light across the earth grew brighter. In April it was light even at midnight. Already the islands stood out clearly in the distance.

In this region, it was around May 17 that the period of what is called the midnight sun began, when the sun no longer sets at all and shines for twenty-four hours on end. At six o'clock in the evening it was in the west, but still fairly high in the sky; at nine, a little lower; at eleven, still sinking, but its blinding light now set the whole sky ablaze. One hour later, it was touching the water. You think it will continue to set, then disappear behind the horizon, but barely has it touched that line than gradually it begins to climb again in full splendor and resume its course – so that its dying rays at midnight are at the same time the first gleams of dawn. At three in the morning, this 'nocturnal' sun is as warm as it is between seven and eight in the morning in Africa.

The oddest thing was that we couldn't get to sleep any more. To fill in the time I stayed at the school, where I took notes, sometimes until three in the morning. Kield Pedersen, the Danish headmaster, kindly gave me access to the *Meddelelser* of his establishment – many bulky volumes which contained the findings of every piece of research done in Greenland since the days of Hans Egede.

Outside, small orange or red tents sprang up, erected by children whom the endless daylight kept from sleeping. At three in the morning you could still see them playing outside. Sometimes they went on like this for two whole days without going to bed. Eventually they dropped with fatigue, and then might sleep for two days at a stretch. It was the teachers, not the parents, who complained, because most of the time their classes were half empty.

Sleep eluded the adults, too. Everyone was restless. They had hardly set foot indoors before they were longing to go out again, to tramp on and on, to run from hill to hill. They rambled around incessantly, in search of who knows what. All through the spring they'd go wandering like this, building cooking fires in the mountains with three stones for an oven, gathering *paarnat* berries, resting no matter where when tiredness overtook them. Both with humans and animals, spring here was the season of tireless frenzies of love. Groups of boys and girls ran laughing and shouting until early morning, and there was the noise of rutting huskies fighting, the deep growls of the males mingling with the bitches' piercing yelps. The birds sang and eiders quacked in the creeks.

The landscape seemed excluded from this general harmony, and it changed overnight. All the filth of Christianshåb was suddenly exposed by the sun's return and the thaw. Snow melted on the slopes, the street became a river of mud, and innumerable streams riddled the ash-grey earth and brought to light piles of old bottles and cans, dog shit, household waste and rotten potatoes. All the garbage which cold and snow had preserved – now swollen with melted water, rotting fast and buzzing with clouds of flies, real flies, come out to haunt us like a bad conscience. Outside the doors and under the foundations, the houses were repulsively filthy. The borrowed coat of spotless white had covered so much offal! A sickening stench hung everywhere. The dogs, some of them now moulting, slunk squalidly about the village. You really wondered whether you were still living in the same land that had once been so clean and white. Flies found their way indoors.

One afternoon Kield's three-year-old son did something very commonplace – he killed a fly. This was all the easier for a child to do because the flies, always less active here than in Africa, seemed even more sluggish in spring. So Kield was unimpressed by his son's achievement and made no fuss about it, much to the amazement of the boy's nurse, a Greenland woman of about thirty, who showed great pride in her charge, precisely because he had killed a fly. 'It's a sign that he'll be a great hunter,' she asserted. If all one needed to become a great hunter was to have slaughtered flies in childhood, then unquestionably I ought to have been one, along with every other African. At the nurse's insistence, Kield and his Danish school teacher wife made quarts of coffee and more cakes than ever before, in order to enlist the whole village in that important occasion, a little boy killing his first fly! The villagers couldn't find words enough to congratulate the child, streamed through Kield's house to see the infant prodigy, and drank coffee until dawn. Surprisingly enough, the hero of this saga was not allowed to touch any of the cakes baked in his honour! And, in fact, in small villages – so the nurse told us, backed up by a chorus of visitors* – an adolescent who has just killed his first seal mustn't eat any of it himself; others eat almost all the meat. During this meal (always followed by a *kaffemik*), which is eaten while the young hunter looks on from a corner, everyone tells him after each mouthful: 'Really, in all my life I've never tasted such delicious seal!' All this is to teach the young man that from now on he must put the community before himself, and share everything with it. What a fine lesson – but what a tough test!

I couldn't help comparing this ritual with the Togo custom which obliges a boy to set some portion of his first wages aside, before any other expenditure, and give it to his father, be it only a token franc. This custom is derived from offering the first fruits of the maize harvest to the gods to protect us from their wrath. But while the Togo youth is thinking only of his father and of cementing his fondness and esteem, the Eskimo boy, through a greater sacrifice, is thinking of the survival of the community. It took the death of a fly to make me realize this!

When he found himself denied a cake, Kield's son began to cry. The nurse took him in her arms and said in consolation: 'Be more modest, little one! Don't keep your first catch for yourself – think of others first!' Apparently, the seal and the fly and cakes were interchangeable in the nurse's imagination. But could the Danish boy have understood, even if he had been older?

From adolescence onwards, a Greenland boy is considered and in fact treated as an adult, with all the responsibility which that implies. Here is another striking example: On May 19 we had the confirmation ceremony in our village for boys and girls

---

* The nurse came from Saattut, a hunting village of 250 inhabitants, located near Uummannaq, some 300 kilometres north of Christianshåb.

of thirteen and fourteen. Among the presents given to Josef's brother Jonas were a rifle and an ammunition pouch, 'for killing ptarmigan'. Fine presents for a boy of thirteen! Where else in the world, even where survival is dependent upon hunting, do people put a rifle and bullets into the hands of a child?

3

Robert Mattaaq

Living through winter and spring in the north of Greenland with only local transportation to depend on meant living in constant uncertainty. I had now spent two and a half months in Christianshåb waiting in vain for a ship to Thule. During this time two cargo boats from the south had reached Thule, but they had kept to the ice-free open sea, avoiding Christianshåb because of the pack-ice lingering in this middle region.

To reopen communications with our village, an ice-breaker had been at work in the bay for the last few days. Laboriously, very slowly, it cut the stubborn pack-ice into big chunks, leaving behind it a vast, unreal field crisscrossed by deep, irregular furrows. But that didn't solve the problem, because these enormous slabs of ice could not be removed from the bay. After all, how could this be done? So after this strenuous labour of partial destruction, we simply had to resign ourselves to wait a few weeks more, until the ice floes broke up more and softened enough to let boats through – provided that an unseasonable return of the cold, always a possibility, did not freeze them back together again. Either way, there was every indication that this May the region would stay cut off for a long time, and would probably have to be supplied by helicopter.*

* There is a process now in common use in Norway to melt the ice in fjords that are also seaways, so as to make them navigable all year round. This pro- cess consists of covering the floor of the fjord with a large number of perf-

In my own case, travelling from Christianshåb to Thule by helicopter was immediately ruled out, not because of my own finances, which would not have stretched so far in any case, but simply because no such link had been set up in the country. And even if it had been, what's the point of flying over a landscape? Wasn't it a hundred times more worthwhile to sail through this natural grandeur, to feel its overwhelming power? Such a feeling was far superior to the superficial admiration experienced from the air! Besides, even though on every sea voyage I make, however short, I suffer from seasickness, a long boat trip is the form of travel I like best. That inner psychological war with the elements which one feels one is waging at sea provides me with a good cure for the indefinable sense of anxiety, or the powerful sexual drive, that the lone foreign traveller often develops so intensely in the idleness of Greenland villages. After all, this inner mental struggle may also amount, by transference, to a sexual act . . .

On June 1, still hoping to find a cargo boat to Thule, I returned to Egedesminde, south of Christianshåb, aboard a small ship. Then came another long month of futile waiting. There was no boat for Thule, but every two weeks or so a vessel left for Upernavik, the second most important town on the northern coast. Tired of waiting, and running short of cash, with deep regret I gave up the idea of a voyage to Thule and decided one fine morning to take the boat to Upernavik.

On July 1, I sailed on the *Tikerak*, a wooden cargo vessel

---

orated pipes attached to suction and force pumps, which carry pressurized warm air under the water. This warm air is forced through the holes in the pipes and forms bubbles which, breaking on the surface, keep the waters of the fjord at a temperature above freezing. These 'aeroducts' – if the reader will forgive my neologism – are an ingenious invention. But the introduction of a similar technique in Greenland is precluded by the extent of the ice sheet and the considerable cost of the operation, which would have to be carried out there over a very wide area.

equipped for ice-breaking, which called at the mining centre of Qullissat the next day, and dropped anchor at Upernavik on Sunday, July 3, at seven in the morning.

About three hundred metres from the landing-stage, the traveller pauses in amazement at the sight of a wooden building with heavy windows and a steeply pitched roof that recall the massive old farmhouses of Jutland. Apparently it was built out of nostalgia for the Danish countryside. This edifice, a kind of giant among pygmies, is all the more impressive here for being the only one of its kind among the little clapboard doll's-house dwellings that make up the village of Upernavik, and our wide-eyed traveller wonders what purpose it could possibly serve. Finally, two Land Rovers parked under its windows, with doors carrying the initials GTO, reveal that the building houses the offices of the Grønlands Tekniske Organization, the state construction agency for Greenland.

The traveller receives yet another surprise when, just behind the GTO building, he spots a cottage made of turf. More than half a century old, and the only one left in Upernavik, it stands alone in the middle of a stretch of wasteland covered with rocks and lichens. Barely two metres high, and occupying two or three times less space than the modern wooden houses, from a distance it simply looks like a mound of earth. Its thick walls and flat roof are made of alternate layers of turf and flat stones, and it is topped by an anachronistic masonry chimney. The entrance to the house is a long, low tunnel through which you have to crawl in order to get to the rooms, so that – rather than a human habitation – one imagines some huge, misshapen fossil creature rising out of the rocky plain, legless but with a long stiff tail.

The presence of this earthen hut in Upernavik comes as a surprise to the traveller who knows that the GTO has been systematically destroying these tough old turf dwellings in villages of more than five hundred inhabitants (here considered 'large townships') and replacing them, in the name of progress, with flimsy wooden houses. Upernavik has a population of almost six hundred. In some of the large villages in the south, where it is relatively warm in summer, one or two big turf dwellings had been spared or restored, because the cool temperature provided by their massive walls made them suitable for storing potatoes . . .

But this particular hut was no storehouse for vegetables, for a faint plume of bluish smoke rose from its chimney, proving that it was still inhabited. By what miracle had it survived the axe of the GTO? Curious, I approached the house.

The turf, which from a distance looked like grey soil, contained grass as dry as chopped straw. Apart from its long entrance corridor and low roof, the house vaguely reminded me of an African mud-walled hut; despite its dilapidated appearance, and probably because it aroused some personal memories, I felt an irresistible desire to share the life of its occupants.

So I opened the door to the entrance tunnel and crouched my way along the two or more metres of corridor. At the other end of this darkish passageway I knocked at a second door and waited for someone to come and open it. Instead, I saw two eyes peering at me through a wide gap above the door. Almost at once the eyes disappeared, and in the same instant a woman's voice let out a scream of terror on the other side of the door.

The woman started calling out to someone and talking volubly. Other voices answered, and there was the sound of hurried footsteps from inside. Finally the door swung open, and I found myself facing a squat little creature whose sex was indeterminate at first sight. It was covered with thick black hair, from which emerged a round face set in a hard mask. Long locks of tangled hair hung down its back; other shorter locks were shoulder-length. Closer inspection revealed that the creature in front of me was an old man, though with quick and lively gestures. His checked shirt was ragged at the elbows, and the tattered sleeves dangled pitifully over forearms plunged to the wrists in the pockets of ample black trousers tied round the waist with string.

His rheumy eyes stared impassively up at me, and he said: 'I've heard about you on the radio since you arrived in the south. I've been waiting to see you for over a year!'

After the usual greetings, he invited me into his hut.

Such was my first meeting with Robert Mattaaq, one of the oldest men in Upernavik, who lived in destitution, aloof from the Inuit, in the village's last turf dwelling.

A wooden platform, a chair, an armchair with two filthy cushions on the seat and back, a rickety table and empty wooden milk crates scavenged from the Danish store constituted the furniture. Two women were seated on the boxes. One of them was old; I learned shortly afterwards that she was Robert's wife. The other was his daughter, aged about twenty, and eight months pregnant at least, judging by her prominent belly. It was she who had peered out at me through the gap above the door; her dark eyes, haggard in her plump, round face, were still wide with the terror she had felt. The ceiling was low – so low that, even at the centre of the only room in the house, I couldn't stand upright without bowing my head.

After waving me to the chair, the old man sank into the armchair, obviously his usual seat. We talked about the journey I had just made, the weather, the state of the pack-ice, and the villages further south.

'*Paniga*,'* he said, addressing his daughter, 'make us some coffee.'

Little by little the atmosphere became friendly. I took this

---

* Term of affection meaning 'my daughter'. The real name of my host's daughter was Bolette.

opportunity to tell my host the true purpose of my visit. But when he learned that I wanted to stay with him, old Mattaaq replied:

'You're too tall. If you live here you'll knock out my ceiling with your head!'

Looking at his broad, tranquil smile, it was impossible to tell if he was serious. This uncertainty lasted for more than half an hour before Robert Mattaaq agreed to put me up. I soon discovered the reason for his change of mind.

Generally, the interiors of turf dwellings were covered with rough wooden panelling to strengthen the earth walls and ceilings. This panelling was made of old planks that for generations had been obtained from driftwood. The occupants used to line this inner retaining wall with skins, the better to insulate their home against the cold. But old Mattaaq had had an original idea: the four wooden walls of his house were lined all over with pages cut out from picture magazines – so many that you couldn't see a scrap of wood behind them! A careless observer might have thought that these pages had been stuck on just anyhow, but far from it. In his own way, old Robert was a 'bookworm' whose favourite reading matter was restricted entirely to periodicals. Every week for many years now he had been getting hold of magazines dealing with 'world affairs'. And even now when he avoided going out as much as possible because of the curiosity his appearance aroused in the village – his wife, his daughter, his youngest son Niels, aged fifteen, and his two other married sons who also lived in Upernavik continued to buy them for him. But therein lay the rub: these magazines, reviews and newspapers began to make such a clutter on the floor that one day old Rebekka suggested throwing them out of the window. Alarmed, the old man began by sorting out this junkheap and pinning on the wall the articles he wanted to reread. And so – casually, almost unintentionally – a first layer of printed pages spread over the four walls, followed in time by a second layer, a third, and even a fourth layer. The ceiling, too high for Robert Mattaaq to reach – and where two sagging planks threatened to collapse at any time – was the only area unpapered. The first pages dated from five years back and, as new pages had kept being added to the old ones, my host had great difficulty locating old articles or documents he needed.

We were drinking our second round of coffee when, learning that I was a French-speaking African, he told me:

'Somewhere in my "library" I have a photo of one of your countrymen.'

Turning to the wall over the bed, he pulled out two rusty drawing pins, examined the pages underneath them, replaced the drawing pins, and pulled out two others further on.

'Rebekka! Do you remember where I pinned up the photo of that big *Qallunaaq* about two years ago?'

'What big *Qallunaaq*?'

'*Frankrigimioq!*'

'How should I know?' his wife replied, sounding rather irritated.

Finally Robert found what he was looking for. You can imagine my stupefaction when, instead of a photograph of a black man, I saw on the yellowed page of some magazine a picture of General de Gaulle! My host cleared a space among the glasses and coffee cups, wiped the table with his shirt-tail, and laid the picture on it. I shall never know whether he took such care in order not to damage his page or out of respect for de Gaulle. The second hypothesis could well be the right one, for the old man said:

'He's an *isumatooq* (someone who thinks a lot). Have you ever seen him in France?'

'Yes, but only from a distance.'

'I hope his health is good?' he solemnly inquired.

'*Aap*, his health is good.'

Robert shook my hand and delightedly agreed to put me up. 'You can stay with me as long as you like! There is *neqi assut* (plenty of food). You will want for nothing!' He himself carried inside the rucksack I had left at the entrance.

From now on, there would be five of us in this hut: Robert, his wife, his daughter, his son Niels and me. Apart from the sleeping platform, there was no sleeping bag and no blanket that I could spread on the floor to sleep on. How were we going to spend the nights? A crucial question, but it didn't seem to worry my hosts, since none of them raised it.

That night at bedtime, Robert was the first to lie down on the platform; he lay on the side nearest the door. A little later Niels joined him and lay on the opposite side, by the wall. I waited, for the women weren't ready yet: they were putting away the dishes, then hanging up clothes on the drying-rack above the stove. They signalled that it was time to sleep: I offered to sleep on the floor, but this led to protests all around.

'No! *Tamatta* (all of us) on the platform – it's warmer that way!' said Robert.

This everlasting need for animal warmth at night is the clinching argument put forward by Greenland hosts when a foreigner declines the invitation to share the bed where they sleep with their wife and children.

So I lay down next to Niels. A few minutes later, old Rebekka lay down beside her husband. Finally, to my great astonishment, Bolette took the space between her mother and me. Such was our order of sleeping for the rest of my stay.

On the first two nights, Bolette slept with her legs drawn up, facing her mother, keeping the same uncomfortable position until morning with a kind of savage obstinacy. Naturally, not being able to sleep forever on the same side, she eventually began to turn over fairly often. Never in my life shall I forget the disturbing contact and effect of her swollen belly, naked and burning hot, pressing against me.

On more than one occasion I happened to go to bed after my hosts. As I read or made notes at the little table by candlelight, I glanced up from time to time at the family all snoring there, half naked, legs intertwined, and wondered if the peculiar sleeping arrangements in Greenland didn't encourage incest. Thinking back over the numerous houses I had stayed in, however, I realized that I had never come across a single case of it. This prohibition, rigorously observed, is remarkable. But for the Eskimo, the restriction goes even further than the family circle: the family taboo forbids any union between collaterals in general; a demographic study of the Thule tribe, carried out in 1952, claims that all marriages between cousins, even cousins six times removed, are forbidden. This doesn't mean that the Eskimo has no innate desire to infringe the taboo. Sometimes transgressions of this type are recorded in the villages. Bolette, for example, didn't know – or claimed not to know – who had fathered the child she carried, yet village gossip alleged that it was her own father. My personal observations gave no justification for paying the slightest heed to this assertion.

Naturally enough, on the common sleeping platform children regularly witness their parents' sexual activities. The reactions of these little ones are varied. Some deliberately start crying to interrupt the act, while others, prevented from sleeping, ask their parents to make less noise. On the other hand, some pay keen attention in the semi-darkness to what is going on and the next day, in the presence of the laughing elders, imitate the pumping hips and moans of parental orgasm, without being labelled as future sexual delinquents. Faced with these expressive mimes, parents have three explanations for their excessive tolerance. In the first place, they would rather have their children imitate what they have seen, like this, without malice. The opposite reaction would be more to be feared, because it would foster dissimulation. After all, it's not their fault – nor for that matter that of their parents – if they witness these acts; the real culprit is the cold, which for generations has compelled the family to sleep huddled together on the same platform. Secondly, these presumably innocent parodies reveal an inclination to make fun of a serious matter. As I had already noticed, the tendency to ridicule people and things is one of the qualities Greenlanders most appreciate. Parents are happy to find this characteristic in their children, even at their own expense. Thirdly – the excuse most often put forward – children grow up, so there's nothing abnormal in their learning the facts of life at an early age.

Still, it is not every child who can watch his parents making love without serious psychic disturbances resulting. I am think-ing of the horrors suffered by the son of one of my hosts in Jakobshavn. Every time G. had been drinking too much, his wife shunned him and the sickening smell of drink and tobacco he gave off. She would go and sleep with her children at the other end of the platform. G. lay on his back with bleary eyes and kept calling out, 'N., come here – I'm waiting!' 'Leave me alone!' she retorted in a sharp, shrill voice. The next hour witnessed an end-less series of appeals followed invariably by refusals. Finally, in his absolute determination for sexual relief, G. would step across his sleepless children to get at his wife. She had her own strange method of putting up greater resistance, which involved hug-ging one of her sons – always the same eight-year-old boy – tight against herself. Protecting his mother by holding her close in his arms, the howling, weeping child would fight off his drunken father with tremendous kicks in the face which sometimes made his father's nose bleed. Soon the man would fall back out of breath, but he quickly returned to the attack amid shrieks and tears from all the children. This unbelievable scene would drag on sometimes till morning, with momentary pauses and savage resumptions. The little boy in question had once announced that some day he would kill his father, because he was making life a torture for his mother.

One evening at ten o'clock, we had just gone to bed when a voice called from outside. Robert and his wife got up at once and hurried down the entranceway to meet the newcomer. It was Elias, one of my hosts' two other sons, who came in carrying a big jute sack of provisions for his parents, as he did regularly once or twice a week. They opened the sack right away and emptied the contents onto the floor in front of the stove – big narwhal ribs with only a few dried shreds of meat remaining, bony reindeer legs, rock-hard frozen birds, and dried and frozen seal intestines. The bones might have been leftovers, but they could still be scraped with a knife and cracked open for their marrow. Although we had eaten already, we all crouched beside the stove, swallowing seal intestines as they thawed, and chattered with Elias until after midnight. It was these bones and intestines that Robert called 'plenty of food' when he asked me to stay.

July 7 was Robert's birthday: he was sixty-three. He got up early and was pleased to find that the floor had been swept clean by his wife the night before. She had also brewed some *immiaq*. Robert washed with a small towel dipped in a basin of water, without removing his trousers or boots. In fact, he very rarely undressed and washed. 'It's warmer if you don't,' he liked to say. After washing, he put on another shirt, just as threadbare as the old one, then settled down to receive the visitors who would soon come for coffee and *immiaq*.

Some of the villagers who dropped in for coffee with my hosts called back the next day to return the invitation. Robert declined, but left his wife and daughter free to accept if they

Robert Mattaaq in his cottage

Robert Mattaaq speaking in riddles with a string, a popular
local game during the long polar winter night

wanted. During my first two months with them, I never saw Robert go out of doors except to walk two or three times around his hut with his hands in his pockets, before coming straight back to his armchair.

Then something happened that did bring a change in his monotonous way of life. On July 23 his daughter Bolette gave birth to a big baby boy. The mother and child returned from the hospital a week later, and after consultation with the pastor, the baptism was set for August 28. On that day, Robert was up and about before dawn. He put on a fine blue anorak, then made his way to the little church on the hill above his house, accepting his countrymen's congratulations on the way. He himself had chosen the given names for the boy: David Hans Johannes Aaron Mattaaq!

'Why "Mattaaq"?' I asked him.

'Because we don't know who the father is, but it doesn't matter anyhow!'

Robert was born in Prøven (Kangersuatsiaq), a village fifty-six kilometres south of Upernavik, where he and his brother Niels, also a hunter, had spent most of their lives. One morning, Niels went hunting seal in his kayak. Night fell, and the villagers were preparing to go to bed, and still he didn't return. They were worried about him, but supposed that he must have run into bad weather and decided to spend the night at the next village.

The following morning, a close relative arrived from the neighbouring village and said that he hadn't seen Niels. A search was organized: boats and kayaks explored the fjords and creeks and scoured the camp sites, without success. They had lost all hope of finding him when, on the afternoon of the fourth day, a hunter returning to the village reported having sighted a capsized kayak drifting at sea. The water was calm but, reluctant to venture out so far, he had come to the village for help. A few picked men ran for their boats and rowed for all they were worth. When they reached the spot, Robert recognized his brother's kayak from twenty metres away, and started trembling, then sobbing. They rowed closer to the kayak and found a horrifying spectacle.

The lifeless body of poor Niels was floating among the slabs of ice, still held inside the kayak by the sealskin apron. His head stuck out of the hood of his anorak, and his hands still clenched the paddle, which curiously enough hadn't sunk. Everything was perfectly stowed in the kayak: the three harpoons, the bladder-buoy, the ice-chisel – everything except Niels's rifle. The Inuit were at a loss to explain this accident, especially since their friend had apparently not killed any game before his mysterious death.

Then suddenly they noticed something odd as they were turning the corpse over to free it: at the front of the kayak were two holes in the skin covering, one in the upper part, the other in the bottom, as if some projectile had pierced the craft. The unlucky hunter's fur trousers were also holed at the same place inside the kayak, and there was a gaping wound in his right leg. This wound and the two holes enabled the Eskimos to devise a plausible explanation for the tragedy. Niels's rifle had gone off accidentally, and the bullet had gone right through the kayak, there at sea, wounding him in the leg as it did so. The paddle still gripped in his hands was evidence of his desperate efforts to get back to shore.

Hauled from the icy water, the corpse turned at once into a bundle of petrified clothes: they wrapped it in a reindeer skin which happened to be on board the rescue boat, and so brought him back to the village.*

---

* This contradicts the travellers' tales which state that the body of a man who dies at sea is simply abandoned. Some of these tales go further in speaking of those who die at home. 'If one of the ancestors of the dead man has

At home, and with some difficulty, Robert recalled, they removed the fur trousers and the anorak which clung to the dead man's skin. Then they dressed him in his best fur garments, for the dead are always buried in their best and cleanest clothes. It was impossible, with their tools, to dig a grave in the permafrost, so the body was laid on the ground at a place not far from the village, and all around the corpse they piled big stones in the form of a rough rectangle. Above these were laid long, flat stones on which they heaped ever smaller stones, so as to give the grave the appearance of a mound, or cairn. At places where the flat stones were not long enough to complete the roof, they made a kind of framework of driftwood and reindeer antlers, which served to hold up the stones placed on top. This was how the dead hunter was buried, but you had only to reach out a hand to touch the body, which was visible in gaps through the ill-fitting stones.*

Among the personal possessions left on Niels's grave were his harpoons and ice chisel, as well as his kayak. No one dared to use it again, even after repairing it, for fear of being drowned, too. The soul of the kayak would take its revenge! For the Eskimos, objects as well as humans possess a kind of soul – a subject I shall return to later. Just as no special ritual is observed in the home before the dead person is carried on the bearers' shoulders

---

perished in a kayak [which was certainly true in the present case], the body of the deceased is cast into the sea, or laid on the shore at low tide to be carried away by high tide; and if the sea is frozen, the body is lowered through a hole made in the ice near the house.' See *Meddelelser om Grønland*, Vol. 10.

Given that almost every Eskimo has lost an ancestor at sea in a kayak, on this evidence one is surprised to find any graves at all in Greenland.

* Today the use of coffins has been introduced into the country so that, instead of the corpse itself, it is now a deal box that shows through the piles of big grey stones. When these coffins eventually fall apart, the huskies rush to the cemetery just as they used to in the old days, to gorge themselves on human flesh often well preserved by the cold.

or by sled to his final resting place, so the tomb itself is neglected, and no one returns there after the burial.

Yet it is important to prevent the dead person's taking offence and plaguing the people of the village, so numerous severe restrictions are imposed on the closest relatives. These apply mainly to the surviving wife or husband. A man who has lost his wife must refrain from hunting or eating meat during the month following her decease. During this period, he must not use any tool whose sound might vex the wandering soul of his late wife. The use of a hammer or saw is for a time forbidden. The poor man must also avoid breaking a meat bone to extract the marrow which Eskimos are so fond of, because breaking a bone makes a noise.

Wives, whom the mourning ritual prohibits from performing some of the most ordinary actions for an entire month, are subject to even stricter proscriptions. A widow must not leave her house or look up at the sky or at the sea, or mention the names of game animals, or eat certain foods. She must not sew or comb her hair in front of a window. She is not allowed to speak except in a whisper, or even to smile, nor is she permitted to throw out the contents of her chamber pot. And, lastly, she remains 'impure' for a whole year.

The period of one month attached to most of these prohibitions (which vary according to locality) derives from the fact that during that time the soul of the deceased is in the mountains, following the rugged path that should lead to its final resting place. Before this journey is over, the soul takes any violation of the taboos as an insult and returns in fury to the village. Whoever infringes the law is responsible for all the misfortunes which then afflict the community, for the injured soul punishes the entire village. Once the salmon disappeared from a lake where a widow had emptied her chamber pot. This transgression deprived the village of fish all spring. There were years

The graveyard in Upernavik, with stones piled over the coffins

A child's coffin seen through the stones

when neglect of the taboos brought famine on the population. Fortunately, the anger of the dead does not always require such cruel revenge: it is sometimes confined to playing a practical joke on the most influential person in the community, the shaman. There is a legend that tells how at one time the island of Disko was much further to the north: one day a powerful *angakkoq* (shaman), intending to rid the region of it, started to push the island towards the far south when, in a coastal village, he noticed a widow leaving her house before the end of the prescribed period of mourning. At the sight of this flagrant offence, the shaman lost his power, and the island stopped where it is today.

Though tradition both in Greenland and my own country burdens women with the harshest mourning regulations, nevertheless it is always their fault – through the violation of those very regulations – when evils descend upon our societies.

The apparition of ghosts on the offshore ice, I thought, must have the same close linkage as in Africa to the flouting of custom.

'Not at all!' old Mattaaq objected. 'A ghost has nothing to do with these customs, and its demands are not the same.'

He explained that the privations of the mourning period are solely to help the soul to make a painless journey to the here-after. It goes in peace when it knows that the living are fully observing the rules of renunciation – proof that the dead person is mourned and missed by the family.

'But that soul is not the one that becomes the ghost. The ghost is the *ateq*' (the soul of the dead person's name).

So strange a belief in the existence of a 'soul of the name', distinct from an ordinary soul, was bound to arouse my curiosity, and I plied my host with questions. I then learned that every individual possesses several souls, six or seven in all . . .

'These souls are tiny little men scattered throughout the

body. Each has his home in a vital organ – the brain, throat, heart, kidneys, and so on – and keeps it going. (For Eskimos the throat, considered the seat of speech, is a vital organ.) If one of these souls, called *tarnit (tarneq* in the singular*)*, falls ill or is stolen by a sorcerer, the organ it looks after becomes sick. The only cure is to find the soul and put it back in place. Only an *angakkoq* can do that. When a man dies, all his *tarnit* escape through his mouth with his last gasp. When I was a little boy, they used to hold a feather to the lips of the dying, and when the down stopped moving they knew the *tarnit* had just left the body. Then there was nothing more to be done: the man was well and truly *toquvoq*, dead. It is then,' my host went on, 'that the *ateq* begins to make trouble. While the *tarnit* have left the body and flown away, the *ateq* clings to it and becomes 'the soul of the corpse and the grave.' It shows itself in the form of a ghost and, depending on the corpse's state of decay, appears in flesh and blood or as a skeleton trailing shreds of rotting flesh. This soul has only one wish: to live again in the body of a new-born child who has been given the dead person's name. As long as no baby has been given that name, the *ateq* will spread terror through the village.

'The ghost is so feared,' old Mattaaq went on to explain, 'that only a short while ago a village would be abandoned after someone had died, or else they destroyed the house where the death had taken place, and those who had been in contact with the corpse would wash themselves and change their clothes. If people wanted to preserve the house, the dead person was taken out through the roof or a window, or else through a new opening specially cut in the wall and immediately sealed up again afterwards. These precautions were to bar the *ateq*'s return to its former home. Even the dead person's name became taboo and was never spoken again – so much so that a survivor with the same name would have to take a new one. If the dead person had been named after an object or an animal, the word for that object or animal was abolished and replaced by another. So our language went through considerable changes, because these transformations were adopted all through the region. A name not passed on to any child was never heard again until the dead person had been forgotten.'

Among the Canadian Eskimos, this passing on must be done in the winter following the death, otherwise the name is transformed into an *agiuqtuq* that causes sickness and death in men and dogs. So if there is no baby available to receive the name of the dead person, 'a newborn puppy will do'.*

There is yet another soul in man which is called *tarningerneq*. Unlike the *tarnit* souls and the *ateq*, it is not localized but conceived as 'a reflection of the man' or his extracorporeal image. During sleep it leaves the body and 'goes travelling'. The dreams which then emerge in the mind of its sleeping owner are simply the projection of the adventures of *tarningerneq*. So the background of all our dreamt exploits is provided by the lands and places, known or unknown, which this wandering soul explores. Its wandering is not without danger for the sleeper, who runs the risk of never waking again if *tarningerneq* is caught on the wing by an evil wizard.

Of all the Eskimo creations in the spirit domain, *tarningerneq* is undoubtedly the most akin to the conception of the soul that we have in Togo. As a vulnerable being, it resembles in every way that 'double' of ourselves which our wizards and witches in Africa capture and consume in their nightly sabbaths. In my country, man is composed of three elements: the body, the double or the soul (*éklan* or *éssé* in other regions) and the vital principle (*louvon*), sometimes also translated as 'shadow'. A man

---

* Duncan Pryde, *Nunaga – Dix ans chez les Esquimaux* (Paris: Calmann-Lévy, 1974).

cannot go on living unless all three elements are together, but his double may very well detach itself while he is sleeping, go wandering about, thus inducing dreams, and then regain its bodily home in order to let the sleeper wake up. Like the Eskimos, we believe that these flights of the soul present grave dangers, for the owls seen near the villages after nightfall – brown plumage blending with the bark of the baobab trees, a strong hooked beak and talons, broad, flat face with huge yellow eyes blazing through the dark – are none other than wizards who have changed their shape, the better to pounce on people's doubles as they leave their huts. By devouring them, they extinguish the lives of those who now possess only the other two elements. But let's get back to the Far North.

Not only do the Eskimos have the peculiarity and merit of having attributed several souls to man, but also, for them, every object is endowed with life: the lamp, the sealskins piled in the loft – they walk around and talk at night. In the eyes of an Eskimo hunter, the Arctic world with its vast, frozen expanses, its barren, snowy peaks and great, bare plateaux – all that drab, white, lifeless immensity of little interest to an African like me – becomes a living world.

Every object, every rock, an iceberg, a big stone, even such notions as sleep or food – each has its *inua* (plural *inue*), its 'owner'. The word derives from the old term *inuk* (plural *inuit*) meaning 'person'. These *inue*, spirits of inanimate things, are not exactly souls but manifestations of the strength and vitality of nature. Even so, they inspire as much fear as human souls. Less than fifty years ago, the inhabitants of the Egedesminde district still used to throw in or put down as an offering a portion of seal blubber at sites and natural obstacles whose *inue* were reputed to be especially fearsome – dangerous precipices, deep ravines, violent whirlpools and so on.

There is an altogether different ceremonial for the souls of animals, especially those whose flesh or skin provides the Eskimo with food or clothing: land or sea mammals, birds and fish. Like man, each of these creatures possesses an individual soul. By killing to stay alive, man exposes himself to the anger of the animal's soul – an anger to be avoided at all costs. So a complicated hunting ritual has evolved to appease the soul of the victim, and the hunter should perform it either before or after the kill, according to the size of the beast. The orca or killer whale, extremely voracious, feared by all other sea mammals – even by other whales – is the most dangerous of marine mammals, just as the polar bear is the most dangerous of the land mammals. When a hunter kills one of these, he shows it the greatest respect.

The hunting of the whale has given rise to the most extraordinary ritual of all. The whale provides the most blubber and meat, and its skeleton represents an invaluable source of materials: its colossal jaws go into the construction of the great turf buildings used for meetings and ceremonies, known as *qassi*; its enormous vertebrae becomes seats; still other bones are employed to make tools and stabbing weapons. Whales are hunted with harpoons which may have either detachable heads or fixed ones – the latter sometimes attached to a long line that carries inflated bladders designed to impede and tire the wounded whale, which will try to escape by diving. But all that ingenuity is not enough to capture Leviathan; men must also prepare themselves spiritually. Before the hunt, they must pray to the goddess of the sea, then purify themselves. As well as in purification by sacrifice, the secret of success lies in abstinence. These great beasts fear the smell given off by women, and a girl is particularly dangerous during her first menstruation. Women must also submit to certain 'magical rites', the aim of which is to attract and imprison the soul of the whale in the igloo.* Furthermore, gifts

---

* James Louis Giddings, *10,000 ans d'histoire arctique* (Paris: Fayard), 293.

are hung from the ceiling so as to greet the whale not as a victim but an honoured guest who bestows a mountain of meat on the village. No hunter may kill an animal before having captured its soul. Since the loss of its soul is as dangerous for animals as for men, the whale whose soul is snared by women becomes an easy prey for the harpooner.

These Eskimo practices recall those of the lion hunters of Niger. Face to face with the wild beast they mean to kill, they address him with long, sustained, rapid incantations meant to cast a spell over him before they shoot their arrows – customs of a secret world made public only a few years ago by an excellent French film by Jean Rouch, *La Chasse au lion à l'arc* (Hunting the Lion with Bow and Arrow). When the wounded lion keeps on roaring, the hunter, chanting his spells at the top of his voice, commands the beast to let himself die quietly, which is just what happens. True, the African hunter's arrows are tipped with poisons which are both virulent and lightning-fast, and the use of these toxic substances would seem at first sight to rule out all comparison with the simple methods of the Greenland hunters. Yet the incantations that fill the plain around the dying lion, and the reverence of the women as they crouch inside the igloo, are proof enough that the African hunter and his distant Eskimo brother adopt the same psychological attitude towards big game. For both of them, the killing of the animal is tantamount to a murder which calls for punishment; and it is to protect themselves against any retaliation on the part of the sea goddess or the guardian spirit of land animals that the crime is masked by all sorts of precautions.

The soul of the murdered whale is celebrated as the sacred guest of the village, and because the festivities marking this event must not be marred by acts of hostility towards other whales, hunting is temporarily suspended. The truce may last from a few days to several weeks or even longer, according to the community's population and their genuine need for meat – for here all principles, even religious ones, give way to hard necessity. At the end of the celebrations the soul departs into the body of a newborn whale-calf. This is how the animal soul performs its transmigration (just like the soul of the name of human beings), and this transmigration persuades the Eskimo that the animal has not been murdered, for its soul has not been destroyed. Hunting may start again shortly after the soul has left the village.

Obviously, the success of the next hunt depends on the treatment given to the soul of the previous victim. The Eskimo therefore takes the greatest pains on its behalf. The harpoon used to kill the animal is carefully placed near the fire so that the soul, still in the head of the harpoon, may warm itself at night. To the soul of his ocean quarry he gives fresh water (or sprinkles a few drops on the muzzle of a dead seal), for living constantly in salt water must give it a thirst! A bird's soul, like a fish's, is given its own special treatment. Such attentions are always well rewarded because, once back in its element, the soul tells of its stay among men and the great welcome it received. Eager to enjoy the same kind of hospitality, the other animals offer themselves to the hunter's harpoon of their own accord.

Robert Mattaaq sat down at the other end of the table and began his tale. I had asked my elderly host to tell me all he knew about the sea goddess, Great Spirit of Ocean, mother of all marine mammals, feared above all others.

The children were perfectly familiar with the story he was about to tell, but they all listened to him with the greatest attention. Old Rebekka was busy washing and counting the empty lemonade bottles she had rescued from Danish dustbins to return at ten cents apiece to the Company, but even she raised her wrinkled face from her task and with her calm voice added an '*Iéh*' of encouragement or a '*Suuuu*' of affirmation to her husband's words.

'Arnaqquassaaq was a girl so beautiful in the eyes of men that all the young hunters dreamed of taking her as his wife. Many asked for her hand in marriage; she turned them all down. The only man who mattered to her was her father, with whom she lived. But one day when he was out hunting, the petrel that dwelled far out to sea came to the encampment in the form of a handsome young man, who carried her off and made her his wife. For several months her father wore mourning, until one day, while he was out hunting in his kayak, he found her again on an island and decided to escape with her. But there was no room for two in a narrow kayak, so he returned alone to the village and went back to the island in an *umiaq*, a big sealskin boat. He waited until the petrel went hunting, then entered the house and urged his daughter to come away with him. They left the island in haste.

'When the petrel came home, he fell into a violent rage on seeing that his wife had deserted him. Swift as the wind, he caught up with the fugitives, skirted the boat, and unleashed a storm. The father took fright and threw his daughter overboard, but she clung desperately to the boat. Then the father seized his knife, and with one blow chopped off the top joints of her fingers. When she grabbed on to the boat again, he sliced off the rest of her fingers. Still she hung on with the bleeding stumps of her hands, so to have done with it he slashed off her hands at the wrists.

'The poor girl sank to the bottom. Her severed fingers became seals and walruses, and she herself the goddess of the sea, the Great Spirit who rules over all sea creatures. Sitting in the deepest depths of the Glacial Arctic Ocean, her legs drawn up beneath her and her head bending forward, Arnaqquassaaq, who now lacked hands, saw to her despair that her long, spreading hair was getting more and more grimy each day, soiled by the sins of men, which turned into filth and stuck to it. The goddess was outraged. As a punishment she gathered all the sea mammals about her and so deprived the coastal villages of game for many weeks. The inhabitants, threatened with starvation, assembled in the great communal meeting house, the *qassi*, to perform the ceremony demanded by the circumstances.'

In Africa, where the expiatory role of blood is important, the sacrifice of a few living victims would have sufficed to make the goddess relent. These victims were usually large domestic animals, but our peoples even went so far as to sacrifice human lives to the sea to ensure successful fishing. I remember that every year in my childhood, with the approach of the season for catching the fish we call *panpan* (which starts around Christmas), our father absolutely forbade us to go out alone at night. We risked being abducted by *kévigan-too*, the 'men with big sacks', who prowled at night, seeking victims for sacrificial offerings. No one in my household had ever taken part in these human sacrifices to the sea, and for a simple reason. I belong to the Watyi tribe, a people of the soil. In Africa each tribe was hermetically sealed off from the rest, and nobody knew anything about the secret rites of a tribe not his own. The fishermen belonged to a different tribe from mine. Their straw huts were strung along the seashore, and we had contact with them only by day. No word of their nocturnal activities ever reached us, and with nightfall the sandy shore between us became as impassable as the ocean. These fishermen selected their victims from among the farming tribes. Some readers may suggest that the 'men with big sacks' were simply bogeys to frighten children, no more real than werewolves in Europe. I would agree but, alas, the same sacrifices have been reported in neighbouring countries . . .

To return to Mattaaq's tale:

'Gathered in the *qassi*, the Greenlanders began their cere-mony. To find out why Arnaqquassaaq was harrowing the community with famine, they didn't try to question the god-dess.' (We would have, in Africa, where many charlatans are past masters in the art of drawing speech from the gods.) 'On the contrary, the villagers meeting in that communal igloo had gone there that evening to confess their sins in public, to find out what broken taboo might have given offence to the 'Woman of Majesty'. Following custom, the ceremony began with the faint sounds of a droning song that all voices then took up in chorus; this chant, recounting past misfortunes, was intended, as the ritual required, to excite the people's feelings.

'Packed in rows around the walls, crouching with their elbows on their knees, they accompanied this plaintive dirge with slow nodding movements of the head. Meanwhile the sha-man spun slowly around in the doorway, holding the flat drum called *qilaat* in his left hand, the stick or *katu* in his right, and beating out the rhythm. These sounds, and their weird rhythm, took possession of the listeners. A woman hid her head behind her neighbour and started to cry. In no time others were doing the same, and soon there were sounds of sobbing from all around the hall. In that stifling atmosphere the villagers began their confessions under the guidance of the *angakkoq*, who in turn lamented, groaned and shrieked loudly. Crushed by guilt for all the taboos his people had violated, the shaman, who had long been in a trance, suddenly crumpled in a motionless heap on the floor. His spirit then left his body and went to the god-dess in the sea.

' Arnaqquassaaq's ocean home was guarded by a huge tail-less dog, but the shaman passed him by without difficulty, because he himself was now under the guidance of his *toornaar-suk*, his guardian spirit. When he reached the goddess, he washed her hair to cleanse it of the sins of men, then combed it carefully and twisted it at the nape of her neck into an elegant chignon.

'After that, the shaman's soul made its way back to the igloo and rejoined the sorcerer's body.

'A few days after this ceremony, the seals reappeared in great numbers, and there was good hunting by kayak once again. Joy and plenty reigned in the camp.

'Ever since that time,' old Mattaaq ended, 'public confessions and the *angakkoq*'s journey to the bottom of the ocean have remained as our people's one salvation whenever a broken taboo or any act harmful to the community has provoked the anger of Arnaqquassaaq.'

The story of the sea goddess was one of the most beautiful that I had the good fortune to hear among those extraordinary men. When it comes to beliefs concerning hunting and fishing, I must admit that Eskimo customs are gentler than those of my own people.

4

'Your Place is Here with Us'

Since the first days of September, the sinking sun of the early afternoon had shed a soft, golden light over Upernavik. Its course, which now lasted barely five hours, was almost horizontal. Its dying rays kindled the frozen waters of the fjords and fell wine-red on the slopes of mountains still in light. Icebergs were calcined ruins in this giant blaze. The purple disc sank little by little below the line of the horizon; for a few moments longer the brindled sky was still painted with wide and luminous streaks, flaming on the face of the waters, turning to violet on the mountain tops, orange or grey above our heads.

These autumn sunsets rarely held much interest for the people of this polar land; not that they were immune to the matchless beauties of their world (some Greenland artists capture the shades of the short Arctic autumn on canvas very well), but the twilight glow which had returned each day for the last two weeks, and which would linger even longer, became irritating to the native. Autumn, as I have said, fostered boredom and idleness here. The feeble rays of the sun, soon to disappear for many long months, still shed a little warmth, but local people preferred their total absence to this painful transition. So it was with a sense of release, towards mid-September, that the village greeted the first snowfalls heralding the long polar night and all kinds of activities on the returning ice.

I awaited this winter, the second since my arrival, with

Robert Mattaaq and family, with his daughter Bolette,
second on the left, and his wife Rebekka

confidence. More than once, the previous winter, I had driven a
dog-sled team alone, perched on my load of frozen fish, often
through starry nights swept by the aurora borealis. In those
moments of intense cold, with my eyes focused on the track
beaten smooth by sleds and my body full of a sense of sweet
well-being, I had never missed my native Africa, for the poetry
of movement on the ice froze up the muggy heat of my native
tropics.

I had adapted so well to Greenland that I believed nothing
could stop me spending the rest of my days there. Apart from a
sled and a dog team, I needed only a fishing boat and a roof of
my own to live happily in the Arctic. Not just anywhere in the
country, though: the southern coast was not for me. I planned
to settle even further north, at Savissivik or Siorapaluk, villages

with marvellous names, in the Thule district. From Upernavik to Savissivik was about four hundred kilometres, which I was preparing to cross by dog sled, intending to make a number of stops which would include Tasiusaq, Kraulshavn (Nuussuaq) or Nutaarmiut, and Kullorsuaq.

But if I were to live out my life in the Arctic, what use would it be to my fellow countrymen, to my native land? Having tried and succeeded in this polar venture, was it not my duty to return to my brothers in Africa and become the 'storyteller' of this glacial land of midnight sun and endless night? After the degradation of slavery and colonization and the struggle for independence, wasn't it the task of educators to open their continent to fresh horizons? Should I not play my small part in that task and help the youth of Africa open their minds to the outside world?

That is why I decided to leave.

The *Vinland Saga* was sailing from Sukkertoppen for Denmark on September 27. I was five days' journey from Sukkertoppen, and to get there I could only wait for the *Umanaq*, the coastal vessel now making its last trip of the year. If an early frost or bad weather prevented it from continuing to Upernavik, its northernmost port of call, my return journey would inevitably be postponed until next summer, in other words for nine or ten months.

The pallid dawn crept timidly through the narrow window, gradually revealing the outlines of the empty hearth, the massive shape of the communal bed with all five of us still in it, the blurred lines of the table melting into the semi-darkness, and finally the curved rim of the privy bucket at the other end of the room.

Mattaaq got up. Hands stuffed into the pockets of the black trousers that he never took off, he shuffled towards the door, relieved himself in the bucket, then came and perched on the edge of the platform where we still lay, and started gently tapping one small dangling foot against the other. We woke and sat up in our turn.

Before lighting the fire, Rebekka, my aged host's wife, picked up a big grimy canvas bag from the floor by the stove and poured some of the contents into a cardboard box in front of us. The bag held dried seal intestines cut up into sticks, which made them look like flat, wrinkled spaghetti. These seal intestines – our breakfast – were tough: it took a lot of chewing to munch them down to a sort of lumpy porridge that gave off a smell like ripe Münster cheese. They were especially prized when eaten with seal blubber, that yellowish, bloodshot fat which had turned my stomach on the day – now so long ago – of my arrival in Greenland, but which now seemed to me a choice side dish for dried meat and fish. Now that I had been sharing these people's lives for sixteen months, their food no longer disgusted me, and I thought nothing of eating a breakfast of seal fat and dried intestines every morning.

Bolette, my host's daughter, sat by me on the platform and smiled, revealing the perfect straight line of her teeth, half worn down to the gums. On her sweet round face, I seemed to read the teasing question: 'How can you love our food so much, and still want to return to your land so far away?' But she remained silent. A little later, she simply asked:

'*Qassinik qaammat* (How many *qaammat*, moons, namely months) will it take you?'

'*Qaammat amerlasuut.*' (Many months.)

Like my outward journey, my return looked like being no simple matter, given the state of my finances; I was sure that years would go by before I saw my native land again.

'But we'd be glad to have you with us always!' old Mattaaq kept telling me. 'We know you. Do you want for anything here? We have everything a man needs – seals and fish in the sea beyond counting. You know that, because you hunt and fish with my sons. You love our food, just like those foreigners who spent the rest of their lives among us. Not many of them, it's true. Every day they talked of going back to their own land, but they never did. They had become real *kalaallit soorlu illit* – like you, true Greenlanders!'

He paused for thought, then went on:

'But I understand you very well. After so many years away from them, you don't know what's become of your own folk, and you want to go back and see them, don't you?'

He may have been right. Do people ever know their true reason for embarking on a long journey? So many causes, motives and impulses intertwine to form the semblance of a reason.

'One day, after many years, you will come back to us,' Mattaaq declared confidently, tugging at a piece of dried seal intestine that his almost toothless gums could scarcely grip. 'I'll be dead and gone by then. But your place is here with us, where you've won everyone's respect.'

'*Suuuu!*' nodded Rebekka, who was equally entitled to speak, but did so only to back up her husband.

For some minutes Mattaaq had been patiently rummaging for something in a big, dusty, lidless wooden chest full of bits of hide, whalebone, harpoon heads and torn, yellowed sheets of paper. He sat on the floor with his head bent and legs spread wide, and slowly sifted these relics, snuffling vigorously from time to time to curb the snot that threatened to fall into the chest, then swiftly wiping his nose on the back of his hand. His thick fingers with their hooked black-rimmed nails picked up a heap of assorted objects which he let fall gently, like a child playing with sand. After a quarter of an hour he unearthed the tooth and claw of a polar bear, the claw still tufted with long, stiff, silvery fur. These were souvenirs of his former exploits, the remains of a polar bear he had killed on the ice 'long, long ago' and preserved as keepsakes. He gave them to his wife and asked her to thread them on a piece of string.

With the sharp end of a bodkin heated red-hot on the stove, Rebekka set to work. The claw was pierced almost without resistance, giving off a puff of smoke and a smell of burnt hair; but it took more than an hour to bore through the tooth! Rebekka next cut a length of string and passed it through the holes. Then Mattaaq took this curious necklace with its two symbolic pendants, came over to me, and tied it around my neck.

'Keep them,' he said. 'I have nothing else to give you for your departure.' He paused. 'The way will be long for you. Be strong and brave like a young *Nanoq*!'

Deeply moved, I thanked him. My own grandfather would have made the same gesture with the same intention, using the trophies of a leopard; but he would have chosen a remote spot and a twilight hour, spoken arcane words, and enlisted all those minute preliminaries and accessories which, by swathing this simple act in mystery, would have given it increased significance. But here, in the land of the great cold, the daily ritual was stripped of that display. Here life was hard, and the pursuit of food more urgent than in the tropics.

The *Umanaq* arrived at half-past eleven that morning.

At Mattaaq's, the house was in an uproar as soon as we woke up – we were starting to pack my luggage. The old floor was covered with objects that several village families had given me during my farewell visits: a bird harpoon with particularly cumbersome accessories; a dog whip; a miniature kayak and sled, completely equipped, with everything to scale; dolls wearing the brilliant women's costume of the west coast; tray cloths embroidered with pearls in geometrical patterns; and finally, gifts from the old people's home, primitive paintings on bits of salvaged packing case. All these presents would go into the big travelling bag, reluctantly emptied of my fine fur clothes; the skins used to make them had not been cured in any way, and although odourless in this icy climate, they would probably decompose in a tropical or even temperate environment.

At noon, I ate my last meal of boiled seal meat with blubber. At three o'clock, we were ready. Mattaaq had put on his best blue anorak, but we knew that the walk to the landing-stage would tire him, so his wife and I tried to talk him out of it.

'At least let me come with you across the waste ground in front of the house,' he said.

I shouldered my rucksack, with the harpoon and the wooden handle of the dog whip sticking out, and carried my suitcase full of books in one hand. Rebekka and Bolette followed us, carrying more packages.

At the end of the waste ground, Mattaaq stopped. I held out my hand.

'No,' he said. 'Go on.'

He watched us walk away. Then he turned around, and I heard him say my name twice. He was crying – the final show of affection from that lovable old man whose death I learned of not long after my return to Copenhagen.

On the landing-stage, where the last boat of the year had brought out a throng of people, I was literally lost in farewells. They helped me put my luggage onto the boat, which bustled with readiness to sail. When I went on board, the villagers still kept waving. From all along the quay came shouts of 'Safe journey, and come back soon!'

The gangway was raised, the mooring ropes thrown on the deck.

Rebekka and her daughter stood silent in the hubbub and turned sad, eloquent eyes on the boat, which was now putting more and more distance between us. Their faces gradually faded.

We rounded the rock. A mist descended on the valley behind us, shrouding the people and the houses, and I couldn't make out even familiar places. Far away, through the curtain of mist that suddenly enveloped the village, an iceberg glinted.

With a heavy heart, I joined the other passengers below.

CLICHY,
SEPTEMBER 28, 1979

Afterword

In July 1981, five months after its publication, this book won the Prix littéraire francophone international, which 'rewards the best work written in the French language by a non-French author on a subject of universal interest'. This exceptional honour for the efforts of a self-taught writer was a distinction that augured well: subsequently, the prestigious *Times Literary Supplement* devoted part of its July 3, 1981, issue to me, with an article by James Kirkup. His glowing review began with a flourish: 'This is surely the most extraordinary book to come out of Black Africa . . .' The full-page article ended with a touching prediction followed by a burst of admiration:

This is the first example of a black man seeking the soul of the Inuit in the eternal snows, and finding his own, thus opening the way for future generations of nonwhite explorers, African and Oriental. This truly outstanding book is one that I literally could not put down until I had, to my great regret, finished it.

As an ultimate, stirring homage, Kirkup translated the book into English. I never had the opportunity to meet this generous and obliging writer, something I greatly regretted on learning of his death a few years ago. What is certain is that I am eternally grateful to him because, in bringing my modest story to a wider audience worldwide, his work paved the way for translations into other languages, most of which followed that same year.*

In Togo, and especially in my village, people claimed that it was the spirits of the ancestors that had brought me back from the land of ice and snow. Their subtle strategy escaped no one when, from my interviews on Radio Lomé and in Togo Presse, my compatriots learned the details of what they now considered a great achievement, and this gave rise to tremendous national pride. But they remained puzzled by my fascination for the villages in Greenland where traditions were kept very much alive, whereas I had fled my own home to escape the demands of traditions that were just as ancestral. It was seen as a trick by the souls of the dead to lure me back home.

But if one single reason for my return were to be attributed to the ancestors, it would unquestionably be the time-honoured ritual in their honour that governs our relationship with alcohol. This involves an entire series of strict practices and ceremonies that play a role in preventing drunkenness. There was no such tradition in Greenland, even though the Greenlanders faithfully observed ancestral rites in all areas of life, particularly whaling. If there are no rituals when it comes to alcohol, it is probably because they do not produce it themselves and so were unaware of the notion, which in our customs is deeply rooted, that this distilled beverage was destined for the gods before being barely inhaled by humans.

When there was a festival in my village, only one person served drinks to all those gathered in the main square. It was often a young man with a loincloth girding his hips who held a bottle of *sodabi*† in his left hand and a glass in his right. Entering

---

* Danish, German, Dutch, Swedish, Italian, Norwegian, Japanese and Spanish by 2015.

† Spirit from distilled palm wine. Literally 'beverage made from So', the suffix of Hêviesso, the god of lightning.

the circle of participants shortly before the sound of drums and other musical instruments started up, he would fill the glass and pass it to the oldest person present. The elder would get up and carry it at arm's length, invoking an entire line of ancestors and divinities whom he invited to take part in the festivities (they would also be served portions of the meal) and, most importantly, ask them to ensure that harmony reigned in the village. On pronouncing each name, he would bend over and pour a little *sodabi* onto the sand at his feet. By the end of his libation, there was only a drop left, with which he moistened his tongue, and then he would give the glass back to the boy who had not moved from his place. The boy would refill it and pass it to the next person, who would do the same, and so on until all those present, men and women, served in the same way, had made the same libations and paid homage to their ancestors. Then the drums and other musical instruments would come on stage for the dancing and singing to begin. The women attuned their voices to respond to the male notes, following the polyrhythm of the instruments without distorting the polyphony of the chorus, so characteristic of our civilization. As a child, I knew that there was only one glass in the entire village,* that alcohol could only be drunk in this way, and that its use was always associated with an important event such as a religious ceremony, a birth, a baby's first appearance outside the hut on its eighth day, the first fruits offered up to the gods at the start of the maize, cassava, yam or sweet-potato harvest, or funerals. And by the end of each of these ceremonies, we were literally intoxicated from the beating of drums, dancing and singing, but not from alcohol. Nothing like that in Greenland, where alcohol-related tragedies made me nostalgic for my country. On the other

---

* Apart from water gourds, some of which the adults used to drink *dé-Ha* (palm-drink) – in other words, non-distilled palm wine – and, in northern Togo, for drinking *tchoucoutou*, or millet beer.

hand, in a society with no written language, the tireless repetition of oral stories benefits the youngest. That was how, thanks to the incessant invocations heard during libations, some of which came back to my mind like a refrain, I became acquainted with my ancestors whom slavery and then colonization had done their utmost deliberately to erase from my memory.

It is probably this priceless gift, both heritage and a testament to my community, that reminded me of my essential duty, which I have written about in this book, to instil in young Africans, in my small way, the importance of opening one's mind to the outside world. Those young Africans who were part of the elite were avid for knowledge in every field. But what about my own family, and in particular my grandfather Séwa Kpomassie, that ageless patriarch who still worked in his field, which he tilled tirelessly with the same hoe that he had acquired before I was born, and which he carried over one weather-beaten shoulder? What about my mother, my aunts and uncles, all illiterate, but who were entitled to hear about my unusual travels even though our language was cruelly lacking in words to describe snow, or ice fields, in other words the Arctic world?

One evening after my return, I tried to explain to my grandfather what snow was:

'*Atayi*,* imagine that all the white birds in the sky are shedding their feathers . . .'

He sat up in his wooden armchair with a cushion to lean on and gazed at me with his bright eyes.

'Tété! In which country have you seen a sky with only white birds?'

'In no country,' I said, gently pressing him. 'Just imagine.'

The next morning, *Atayi* made the entire family laugh by telling them that in his sleep that night he had seen snow falling.

* Word meaning 'respectable grandfather', because we didn't call him by his first name.

The snowflakes were so thick, he claimed, that they masked the sun!

An uncle who was a hunter asked him what had happened to all those birds that had lost their feathers. 'They no longer have wings so they can't fly away, you can gather them up with your hands!' And everyone laughed all the harder.

What did it matter! But for the first time in the existence of our village below the Equator, snow had entered into the community's conversation.

Better. Before I left Togo, tradition had it that even though I was an adolescent, I should be present at the evening discussions between the adults gathered in my grandfather's sitting room. I was to listen to them without taking part in the deliberations. But now, in an inconceivable reversal of values, grandfather, my father and my uncles and aunts were listening to *me*, not allowing themselves to interrupt me. They had given me life, but, contrary to their expectations, Greenland made me into a man before my time, like an Inuk in the heart of Africa, almost a sage in the African sense of the word, and this at the age of twenty-seven . . .

But after returning to Togo in January 1969, I stayed among my people for only a few months because, in May that year, I set off again with my rucksack on my back on a non-stop, two-year personal tour of sixteen countries in Western and Equatorial Africa.* Travelling by bush taxi and on rickety trains, I stayed with local people, my only resources being the paltry fees I earned from giving talks about Greenland in schools and in French and local cultural centres. That tour gave me the satisfaction of having achieved, with my modest means, the goal I

---

* Togo, Ghana, Ivory Coast, Liberia, Mali, Senegal, Upper-Volta (Burkina Faso), Niger, Nigeria, Cameroon, Chad, the Central African Republic, Congo-Brazzaville, Gabon, Zaïre (Democratic Republic of Congo), Dahomey (Benin).

had set myself with regard to young Black Africans. Ten years later, that trip was followed by a second tour of those same countries, from October to December 1981, organized this time by the Association for Artistic and Cultural Development, ADEAC, under the auspices of the Alliance Française, the French Co-operation ministry and Éditions Flammarion, to promote my book. Excerpts soon appeared in French textbooks in schools in Africa and Madagascar, attesting to the interest sparked by those two lecture tours.

Enabling young people of Black Africa to learn about Greenland and the life of the Inuits through my story is my small contribution to opening their minds to the outside world. But I still felt that I had only partially fulfilled this ambition, since I had not been able to give talks in East Africa, which I still dream about, or in South Africa where, solely because I am Black, I was barred entry due to the brutal apartheid ideology, shamelessly institutionalized by a despicable White minority who massacred the Black population and brazenly defied the opprobrium of the international community.

Apart from Ghana, which gained independence in 1957 thanks to the courage of its remarkable liberator, Kwame Nkrumah, all the countries I visited, including my own, were still under colonial rule when I embarked on my Arctic journey. Since I was the only African to travel the world at that time, I saw myself as the first to emancipate himself from the twin stranglehold of tradition and the occupier. What is more, finding myself in all those countries that exercised self-determination, I was filled with a double sense of deliverance, discovering at last that highly sociable, humane Africa which the colonizers had concealed and skilfully misrepresented as being uncivilized so as to further the so-called edifying histories of their own countries. Then, I was suddenly struck by the irony of my presence in those places, when having run away at the age of sixteen and a half, swearing never to go back into the sacred forest, I had had no intention of returning. So, what had happened?

Yes, it is true that one day I decided on the spur of the moment to go back to my homeland, whereas I was no more than a stone's throw from Thule (Qaanaaq), the region where I could have met the Polar Inuit, Greenland's only isolated tribe living in extreme conditions, and this homecoming was never going to be easy. It was September and the cold was rearing its head after the brief autumn; the sea was starting to freeze over in the fjords, in the channels and on the shore, making the arrival of the boats that still ventured this far north to Upernavik unpredictable. The branch of the KGH (Den Kongelige Grønlandske Handel), the Danish trading post, was no longer able to provide precise, reliable information on these arrivals. The manager, Mr Søltoft, didn't know what to tell me, then eventually he said, 'I don't have any further information about these boats' itineraries. I haven't received the new timetable yet [that of September 1966 to September 1967] and I haven't listened to the radio. It all depends on the weather over the next few days . . .'

To cap it all, the *Tikeraaq*, on which I'd planned to sail to Thule at exactly that time of year, had been destroyed at sea the month before, by a fire. One thing was certain: the last boat that could take me back to Denmark that year was the *Vinland Saga*, which so far had not suffered any damage. It was due to set sail on September 27 from Maniitsoq, much farther south, so I would have to return there in order to embark.

I also had the option of waiting until the middle of winter to make the journey to Thule by dog sled, across the frozen Melville Bay, but the whole of Upernavik advised me against it since it was too risky for one man on his own, and I would have needed some twenty sturdy huskies – even more including replacement dogs – and would have had to hunt every day to feed them. And the mandatory stops at a large number of settlements along the way, most of them located in the sinuous labyrinths formed by the crevices of this undulating rocky coast, would at least double the distance.

To go back to Denmark, my only option was to head south and take the last boat of the year before the sea became an ice field once more, cutting Greenland off from the rest of the world for six to eight months.

I left Upernavik on Saturday September 10, 1966, at 4 P.M. on board the coastal vessel *Umanaq*, which was making its last voyage of the year to that remote town. During the five-day voyage, it stopped to drop off passengers and pick up others in the towns where I had stayed the previous year.

At each port, the inhabitants, so reserved when we had first met sixteen months earlier, shouted for joy and called out to me. My former hosts marched determinedly up the gangway, embracing me, grabbing my rucksack and dragging me back to their homes (*'illumut!'*), thinking that I was coming back to live there.

But Erik Rasmussen, my former host in Maniitsoq, father of little Alfred seen in the photo with me (see p. 149), caught me at the right moment and took me directly to his house. Since my departure a year earlier, there had been a new addition to the family with the birth of a ninth child, born on April 12. But because one of their daughters, thirteen-year-old Emilie, had gone to Denmark for three months in July, and one of their sons had left that day on the *Umanaq* to go and study in Nuuk, the capital, the family had the space to put me up, just as in the past.

'Alfred is always asking for you,' said his mother, greeting me.

I felt all the more welcome because the aurora borealis, which I'd seen for the first time here, was constantly deploying its voluminous curtains shimmering with a thousand colours during this new twelve-day stopover. I spent the time net fishing with Erik – who had bought a motorboat – for *kapisilik*, Atlantic salmon, or *salmo salar*, wild salmon if ever there was any, whose exquisite, delicate flavour we enjoyed at almost every meal.

This dish, served with rice and a saffron sauce, was simple and good. Even though the captain of the *Vinland Saga* had spontaneously invited me to move into my cabin from September 22, five days before the departure for Denmark, if I so wished – and I accepted, taking on board my suitcase full of books – I carried on eating and sleeping with Erik's family.

We set sail as planned on Tuesday September 27 at 5 A.M., with a two-hour stopover a week later in Thorshavn, on the Faroe Islands, to drop off a few passengers. It was the middle of the night, but silhouetted against the sky were the mountains at the foot of which the Vikings had come ashore before Iceland, on their way to Greenland, or Vinland ('land of vines'), part of a continent also discovered by them, and which would later be called . . . America.

On my arrival in Copenhagen on October 8, I was put up by the family of a Danish friend of mine who lived in Vedbæk, a residential suburb, and I spent entire days visiting the galleries in the national museum devoted to the collections of the North American peoples. On October 16, I took a night train for Paris.

When I arrived at the Gare du Nord at 16.50 the following day, I hired a porter to take my bags to the left-luggage office, where the registration process took some time.

Then, exiting the station, I took a long walk, wandering aimlessly up and down the boulevards, avenues and streets, with no specific destination for once. The deafening roar of traffic would disorientate a sled driver and panic his dogs, accustomed to the barely audible sliding of runners on the snow, the gentle scratching of the dogs' claws on the ice or the cracking and dull but brief rumble of an iceberg turning around in the frozen fjord. I had left all that for an infernal world. I found a semblance of calm on reaching the area around the Saint-Martin canal, but I realized it was time to go home to my adoptive father Jean Callault in rue Philibert-Delorme, and I took a taxi.

When I rang the bell, the concierges (a very obliging couple), gave me a rapturous welcome, pleasantly surprised to see me in excellent shape, living proof that, contrary to what they believed, 'the Eskimos aren't savages who eat all their food raw'. They had been kept regularly informed of my adventurous travels in the frozen north and counted the letters I'd sent from Greenland. The sixty-second had arrived on Saturday, two days earlier, against my father's forty-second, which had been brought to me on the boat the very day of my departure from Upernavik. We numbered our letters, at my suggestion, so as to be sure they had arrived safely, mine containing excerpts from my notes so they would not be lost should my diary come to grief.

After a bath, I settled into my room – actually the study with a camp bed made up during the day by the cleaner.

The after-dinner conversation on the evening of my homecoming went on until two o'clock in the morning, because the inexhaustible account of my travels stimulated an exchange that was all the more fascinating now that my father knew Greenland like the back of his hand.

'I couldn't have discovered it more fully any other way,' he said, rising to wish me good night. He added, 'As you can see, I'm finding it harder and harder to walk. But how I travelled through your letters!'

'And me through your generosity. Thank you again for your precious support and all those postal orders!'

'Come, come . . .' he concluded, 'my small contribution to your life project is very little compared with your own efforts. But now you need to rest . . . we'll talk more tomorrow.'

When he awoke at six o'clock, as usual, he saw a light under my door and softly pushed it open. I was already up and writing about the last part of my journey from Copenhagen in my notebook.

That day, I collected my bags from left luggage and settled into rue Théodore-de-Banville, where my father had found me lodgings, planning among other things my return journey to my homeland. I racked my brains trying to devise a roundabout way of describing Greenland to my family in our Mina language, which I hadn't heard spoken for twelve years, and which had no words to explain that universe.

I was mulling this over when, one afternoon in March 1967, after a long walk to avenue de Wagram, I entered a brasserie, Le Petit Wagram, and ordered a beer at the counter. I struck up a conversation with another customer, who turned out to be Gilles Sala, a West Indian musician whom I knew by name and whose columns on Caribbean and African music I'd already read in magazines such as *Bingo*, where a few years earlier I had also published a humorous story which he said he'd enjoyed. Literally petrified by the account of my life as a seal hunter on Greenland's ice floes, he suggested we move to a table, where we carried on talking until it was time for me to go to my father's for dinner. He gave me his card and wrote down my address.

To my amazement, the next day I received a telegram from Catherine Bailly of the Office for Radio Cooperation, OCORA. Gilles Sala had told her about our conversation of the previous day and passed on my address. A producer, she would like to devote a radio programme to my life in Greenland. The interview was conducted on March 28 and broadcast straight away across French-speaking Africa. Listeners responded enthusiastically, and I gave further interviews to the OCORA, including one that was televised. I was also invited by the BBC, the Deutsche Welle and the Voice of America, which brought my story to the attention of the whole of sub-Saharan Africa. And that was how my family too learned of it through the media.

The print press was not to be outdone. Wanting to give his readers an exclusive preview of my travel writings, Olympe Bhêly-Quenum, the editor of *L'Afrique Actuelle*, a monthly magazine published in Paris and aimed, like *Jeune Afrique*, at the continent's elite, commissioned me to write a monthly article several pages in length.

These pieces came to the notice of experts not only on Africa but also on the Arctic. The first was Robert Cornevin, former colonial administrator of Togo and Dahomey (Benin), author of a *History of Togoland*. Head of the Africa and Overseas Territories Research and Documentation Centre, he received me in his office on quai Voltaire.

A journalist who had read my articles contacted my father, seeking to put me in touch with Jean Malaurie,* whom I met initially just before I went back to Togo, then again when I returned to France after my first African lecture tour. Jean Malaurie welcomed me with open arms at his Arctic research centre, and I was a frequent guest at his home in Paris and at his country house near Rouen. Our mutual experiences gave us endless topics of conversation for years, including traditional life in Togo and in Greenland. He was the obvious person to write the Preface to my book.

In that same period, Dr Robert Gessain, the author of the precious book that had inspired my Arctic adventure, who was also interested in my Greenland diary, gave me the warmest welcome at his family home in the sixteenth *arrondissement*, and then at the Musée de l'Homme, of which he was the director and where I had the ear and respect of his colleagues. Incredible as this may seem, he recommended my manuscript to Éditions Flammarion . . . And how can I not mention here my emotional

---

* (*Translator's note*) Jean Malaurie: cultural anthropologist, explorer, geographer, physicist and writer. He and the Inuit Kutsikitsoq were the first two men to reach the geomagnetic North Pole, on May 29, 1951.

meetings with the explorer and ethnologist Paul-Émile Victor? These often took place at the Maison du Danemark, on the Champs-Élysées, where I was invited to all the cultural events devoted to Greenland and the Arctic.

Newspapers in the African capitals published excerpts from my writings, which the radio stations broadcast in local languages aimed at the villages, all thus paving the way for the return of the prodigal son.

I left Paris on Thursday December 26, 1968. On Christmas Day, I had eaten lunch and dinner at my father's. He was worried about whether I would be able to adjust easily to the tropical heat. 'You will melt like snow in the sun,' he predicted.

We both dissolved into tears when the time came to say goodbye, at around nine o'clock that evening, crying and vigorously shaking hands as we faced each other across the kitchen table where we had just eaten a hasty farewell meal.

I had already taken my 120 kilos of baggage to the Gare de Lyon during the day, and, because my father had difficulty walking, we said our goodbyes in the doorway of his apartment. There was a lengthy embrace, outpourings, and more entreaties to take care, followed shortly afterwards by a final wave of the hand from the window, which went on until the taxi taking me to the station had turned the corner on to boulevard Berthier. More overwhelming than the pangs I had experienced so many times on my various leave-takings, it was truly heartrending to leave my adoptive father, the most humane person I have ever met. It was as if destiny had placed in my path, just when I needed it, in the form of postal orders that represented great sacrifices for him, the money I needed for my Greenland voyages and which I could not earn myself, since I was not permitted to work in Greenland.

My departure grieved him too because, on January 3, when the ship I boarded in Marseille the next day stopped over in

Dakar, I received a first letter from my father telling me how sad he felt.

After stopping in Dakar and then Abidjan, the vessel sailed along the coast of Togo on Monday January 6, 1969, in the late afternoon, dropping anchor in Cotonou at seven o'clock the next morning.

If the twelve-day voyage had given me the impression of inexorably advancing towards an inferno as we entered the torrid tropical region, the chill early-morning wind and the relative coolness that reigned for a brief moment when we arrived were still not enough to prevent my shirt from being soaked through, which prefigured the ordeal I would have to face in the coming days.

I hired a porter to carry my luggage off the ship and out of the port once I'd cleared customs. I did not set out for Lomé in a bush taxi until mid-afternoon, at the hour when the setting sun had stopped melting the tarmac on the roads, reducing the number of punctures and other accidents. But first, I ate lunch in a small restaurant in the port, a dish of rice and fish, my first meal on African soil for so many years! The spicy food did not stem my perspiration but made it worse and made me think of the halibut that I ate raw on the ice floe, which was inconceivable for the customers of this restaurant when you think how quickly fish barely out of the water begins to go off and smell bad in our hot climes.

The bush taxi entered Lomé at the end of the day and followed my directions to our Kpéhénou neighbourhood. The district had changed beyond recognition. Everywhere concrete walls had replaced the former enclosures of shrubs, branches, trellises or doormats that surrounded the patios, whereas inside the courtyards, galvanized metal sheeting, some of it rusty, gleamed in the place of the thatch and straw roofs that had once covered the huts and other interior buildings. Not tarmacked, but of brick-red laterite, the new roads were much wider than the former sandy paths and passages that wound between the dwellings and the coconut palms. To achieve this striking transformation, it had clearly been necessary to fell not only the coconut palms but also all the mango trees, bushy and imposing evergreens, in which, perched on the sturdy big branches completely hidden by the dense foliage, my brothers and I used to eat much more than our daily fruit ration by simply reaching out to pick the mangoes that were already ripe and gorging ourselves on them, before they were pierced and damaged by the beaks of greedy birds. Thanks to the stones, picked clean, that we hurled like missiles, the mango trees propagated by themselves, like other plants in our surroundings. Now stripped of its magnificent trees, which, as well as being sources of food, provided much-needed shade, the neighbourhood looked bare, unrecognizable, and at the mercy of the sun's cruel rays during the day. As if modernization had to entail the destruction of our heritage in every area! Without the paltry earnings I used to make from selling the branches and fibre of the coconut palms, so thoughtlessly destroyed, I would not have been able to buy the books I read, or perhaps ever to have discovered Greenland.

As if in a foreign country, I had to ask for directions, and neighbours guided me to our house. I opened the door. If a stranger stepped over the threshold of a house, they would announce their presence saying, '*Agoooo . . .*' (a term asking permission to enter). The occupants responded with '*Amee!*' – let a human being enter, meaning 'be the bringer of good news'. Only the members of the family, in their comings and goings, were exempt from declaring their arrival on stepping on to the patio. So, I went in without signalling my presence and started determinedly crossing the courtyard, where everyone was quietly going about their business.

Nagan, a respectful term we used to address the first of my father's eight wives, whose real name was Gbalessou, was the first to spot me. She rushed over, spoke my name and threw herself at me shouting, 'It's Tété! It's my son! *Tété wézon!*'* because each of my father's wives considered and treated the children of her co-wives as her own.

My mother raced from her hut and embraced me with every ounce of strength in her arms, also shouting, '*Wézon, Tété!* My son! My precious boy!'

She was both crying and laughing, thanking Mawou (the Only God), our supreme God, who, although invisible, at that moment was answering all her humble, lengthy entreaties by bringing back from the unimaginable 'land of snow and ice' her son, whom she thought she would never see again. She looked at me and asked, incredulously, 'They say that over there you ate raw fish?'

'Yes, *Dada*.'†

All my *navi* came running, followed by my brothers and sisters and nephews and nieces who had been born during my absence. The joyful shouts also brought in the neighbours, who had listened to my story on the radio.

I hadn't told my family I was coming back so as to avoid the excitement that would have resulted in several people travelling by bush taxi to come and meet me when I disembarked in Cotonou.

Beaming, my father elbowed his way through to me in his own home and embraced me, saying simply, '*Wézon!*' At that moment, I understood that he had forgiven me and wiped from his memory the unpleasant recollection of my running away,

* 'Welcome Tété'. From *Wé*, personal subject pronoun for the second person singular, and *zon*, the fact of having walked. Literally 'Thank you for [or congratulations on] having come from so far away'. The plural, for several people, is *miawézon*.

† 'Mother' – word used when talking directly to one's mother.

buried under the pride he now felt on hearing our family name on the radio, on reading it in the newspapers and, above all, when people stopped him in Lomé to congratulate him on the exploits of his 'heroic son'.

That night, despite my protestations, he let me sleep in his hut, which comprised a bedroom and a living room. He had my brothers carry his belongings to the hut of one of his wives. He wanted to install an air-conditioner for me, which I adamantly refused. He foresaw that, given all the compliments he received in town before my arrival, lots of people eager to meet me would flock to his home and he wanted me to receive them in comfortable surroundings. That is indeed what happened when Radio Lomé and *Togo-Presse* announced that I had come back and began broadcasting and publishing new interviews. The family home was constantly filled with avid visitors. There were primary-school pupils brought by their teachers, some of whom had once taught me, which delighted my father and flattered his pride; lecturers and students from l'Université du Bénin* in Lomé (Togo), as well as high-school students. We arranged for me to visit their schools. The President of Togo, General Éti-enne Gnassingbé Eyadéma, granted me an audience, which I remember clearly. I was also received twice by His Excellency the French ambassador who, to express his admiration, asked his cultural adviser to organize the lecture that I gave at the French Cultural Centre in Lomé.

At the same time, I had the idea of self-publishing extensive excerpts from the 660 pages of the diary that I had meticulously kept in Greenland, and I applied to the Public Prosecutor's office for authorization to set up a newspaper. I then found out

---

* (*Translator's note*) L'Université du Bénin in Lomé was named not after present-day Benin, formerly Dahomey, but after the Gulf of Benin, sur-rounded by Ghana, Togo, Dahomey and Nigeria and which is itself part of the larger Gulf of Guinea.

that, to exercise this right, you needed to have at least a BEPC (Brevet d'études du premier cycle),* which I did not. But one of my younger brothers, Têtêvi Émélo Étienne, a primary-school teacher posted to Togo's central region, had attained the requisite qualification before gaining his baccalaureate. So as to comply with the regulations, I made him editor of my newspaper. Then I had to convert my father's modest living room into a printing works. The money he wanted to spend on an air-conditioner was used to purchase, from Nopato (Nouvelle papeterie togolaise), the necessary equipment to produce a newsletter: a manually operated Roneo machine,† ink to coat the drum, stencils on which I typed, or rather punched out my text on a typewriter and which, wrapped around the drum, served as a stencil for making duplicate copies, and finally, reams and reams of paper and a big stapler. We printed several hundred copies of each issue. My brothers and sisters pitched in to help with the printing and layout, while I took on the role of editor-in-chief as well as that of courier, because I personally delivered and sold the paper in Lomé, where it enjoyed a huge success, especially among the ministries, embassies and businesses, and I was soon able to reimburse my father.

I could have carried on and established a family newspaper publishing business, diversified the content of my magazine by including writing by young African authors and perhaps helped found a publishing house that was lacking in sub-Saharan Africa. An iron will, strengthened by unwavering family support, made me even more eager to embark on this new adventure, while my mother, who was the third wife, my two *nagan* and my five *navi*, not to mention my sisters, did their utmost to cook delicious local dishes which I had been deprived of for years, secretly hoping that they would put me off wanting to go travelling

* School certificate at the end of four years of high school.
† Machine for duplicating a typed text using stencils.

again and above all that the audience with the President and the warm welcome I received in the ministries would lead to some position that would keep me in Togo for ever. But the resounding success of my lecture at the French Cultural Centre in Lomé, promoted by the media, kicked off my first African tour.

I had the bright idea of taking with me the stencils containing the many articles I had already written, because at every stage, the French cultural centres used them to reprint free of charge the brochures that were sold during my talks, and I should like once again to express my gratitude to them for this invaluable service.

Extensive excerpts from these articles, constantly revised, were to form part of the manuscript of this book, proving that I had intended to write about my experience prior to my encounter with Jean Malaurie. But he persuaded me to write a book in the first person instead of my planned impersonal novel about the life of the Inuit hunters as described in my Greenland notebooks, which he had read. This was wise advice from the publisher of *Terre Humaine*, the prestigious series dedicated to authentic first-person accounts, but where the slot available for a book on Greenland was already filled by his own book.

My first-person account came close to winning the Thomas Cook Travel Book Award in 1983, because I was one of three finalists out of more than one hundred entries. In the end, my book came second, after *From Heaven Lake* by Vikram Seth about his travels in the Himalayas, to the great disappointment of my British publisher, Secker & Warburg, who believed wholeheartedly in the unquestionable originality of my adventure. But the honour of being one of the three finalists was reward enough in itself, because travel guides series such as Lonely Planet, the Guide du Routard and many others referred to my book and thus helped it become known internationally. Once again, I found myself travelling the world, only this time the huge difference was that I no longer had to worry about my travel or accommodation costs.

Still yesterday, many around the globe vociferously claimed that there was nothing of value in my ancestral traditions . . . but I acknowledge that my distant ancestors who instigated snake worship (in that they were more respectful of nature than those who came and massacred those pythons to make handbags), had so to speak put in place the rites and prepared the ground for the initiation journey which, from the child coconut-gatherer that I was in the tropics, turned me into a young Arctic fisherman and into the perpetual traveller that I have become.

How then could I reject the beliefs of my ancestors and betray them in favour of those who, more adept than estate agents, sell us in exchange for gold and diamonds a dwelling in paradise whose door we will only enter once we have starved to death here on Earth? But we have known worse with the same! In the heart of the Arctic, I was unpleasantly surprised and shocked to learn that the peaceful little town of Upernavik, where my crazy polar adventure ended and where I spent a memorable couple of months with my old host Robert Mattaaq and his family, had given its name to one of the Danish slave ships in the seventeenth century.* Through this name, Denmark, one of the very first European countries to organize the slave trade following the Arabs' barbaric raids on the populations of sub-Saharan Africa, was honouring its trade agreement with Greenland. Why hunt us down relentlessly since the seventh century in our own countries, sell us in all the Islamic societies, then, for more than four hundred years in the marketplace across the Atlantic, us, but never the Inuits, even though they were geographically closer to the Danish West Indies in the Caribbean? A commercial choice to plunder that makes me feel even more bitter given that my country, like

---

* Thorkild Hansen, *Slavernes Skibe* (*Ships of Slaves*) (Copenhagen: Gyldendal, 1968); the first book in a trilogy by Hansen on the Danish slave trade.

neighbouring Ghana (then called the Gold Coast), is located right at the heart of this 'slave coast' where the European ships were still continuously loading their cargoes of 'ebony' one century before I was born. My siblings and I could have been part of that cargo had we arrived just one hundred years earlier in this beautiful world which, contrary to all expectations and spurred by my ancestral beliefs, I continually travel today as I please. Vodou, the strength and watchword of the struggle for independence which won back the freedom that had been taken from me as a result of greed and with the help of foreign religions, deserves all my devotion.

My return to my home country ended up being for a very brief time, much to the despair of my family, but they were consoled by the frequency with which I now kept in touch. This has been much easier since I moved to France, where I have started a family. I have taken them to visit my native village in Togo several times.

But I have also made three more trips to Greenland, the first with my family in the summer of 1985, to Disko Bay, Sisimiut, Ilulissat and Qasigiannguit, where in each place we stayed with my former hosts; the second in the winter of 1988 for the shooting of the BBC documentary (*The African Eskimo*) about me, and the last for three months during the summer of 2007, where I gave talks on board the *Fram*, a Norwegian cruise ship plying the west coast.

The publication of this book not only led to a second lecture tour in several African countries, but it also meant I was often invited to the Netherlands, Great Britain, Denmark and Norway, and on two unforgettable trips to New York, first of all in Febeuary 2003 to promote the NYRB edition and to give a talk at the very prestigious Explorers' Club, and then for the opening of an exhibition organized by Jean Berberis, curator of Flux Factory (FF), in cooperation with the Arctic Book Club (ABC), and Michelle Levy, Program Director of the Elisabeth Foundation

for the Arts (EFA), of the works of some twenty young paint-
ers inspired by my book, a show held at the EFA Project Space
in Manhattan from September 18 to October 24, 2009, under
the title 'Artists Respond to *An African in Greenland*'.

Many authors would have liked to be in my shoes during the
various Norwegian literary festivals where I was a guest speaker –
in Oslo, Bergen, Stavanger, Trondheim, Tromsø and, more
recently, Finse. These trips and many others, some of them under
my own steam, have taken me to all the Sámi countries* in the far
north of Norway, in other words the entire Norwegian part of
that vast northern region known as Finnmark that extends over
three neighbouring countries: Finland, Sweden and Russia.

I should like here to express my profound gratitude to the
Norwegian government which, through its Ministry of Foreign
Affairs, accorded me the very great honour of inviting me in
September 2007, along with eleven other international authors
who have written about the polar regions, to visit the 'Norwegian Far North' in order to see the progress that has been
made by the country. Dubbed Ultima Thule, this journey enabled me to discover Spitzbergen, and to revisit Tromsø and
Hammerfest before going to Kirkenes and venturing across the
border from the locality of Grense Jakobselv to Russia. During
my stay on the island of Spitzbergen, at the library of Long-
yearbyen, the capital, I signed the copy of my book which the
inhabitants passed on to each other. I was moved when the
mayor, awarding me the insignia of his town in a presentation
case, offered me free accommodation if I decided to stay on this
other island at the end of the earth 'to write another book'.

The episodes I have recounted so far remind me of the compliments my Norwegian publisher and dearest friend Asbjørn
Øverås of Aschehoug & Co. Publishers (driven by the desire to

---

* Tromsø, Hammerfest, Nordkapp, Honningsvåg, Alta, Karasjok and
Kautokeino.

know more about the local culture, he took a trip to Benin and Togo in February 2017 and visited my family in Lomé) paid me one day when he said that he knew of no other author who had travelled as much as I had after writing just one book! So, it is not hard to imagine that my return to Togo was not a homecoming in the strict sense of the word. To those who might still be wondering whether I plan to go back for good one day, I don't think my feelings on the subject have ever changed:

Since the previous June, I had covered the whole south of Greenland, then Disko Bay which can be considered the borderland between the South and the Far North, and I had come to know the people and customs of the various localities. Now I felt a longing to go to Thule, the most northerly 'town' in Greenland and the whole world, and spend the rest of my life among the real Eskimos – an inexplicable, muddled, yet vivid wish, triggered perhaps by the pursuit of a recurrent dream, the lure of the distant unknown, or quite simply by the fatigue caused by constant travel and a desire to find some last fixed point which would be neither southern Greenland nor Africa, and above all not Europe!

In my heart of hearts, I don't think I ever left Greenland.

NANTERRE,
AUGUST 18, 2014

Glossary

Note: 'Mina' is one of the languages of Togoland; 'Inuit' is used here to denote the language of the Greenland Eskimos.

aap (Inuit) yes
akassa (Mina) a porridge with a maize flour base
ammassat (Inuit) caplins, small dried herrings
angakkoq (Inuit) shaman
asavakkit (Inuit) I love you
Atavi (Mina) paternal uncle

canari (French) earthenware jar

eqalussuaq (Inuit) blue shark

Fofo (Mina) honorific title for all elder brothers; also used by a
    woman addressing her husband
Fofogan (Mina) honorific title for the eldest son of the father's
    first wife
Forsamlinghus (Danish) town hall

Hêviesso (Mina) the god of lightning

Ilulissat (Inuit) the settlement of Jakobshavn
immaqa (Inuit) perhaps
immiaq (Inuit) Greenland beer
Inuk, plural *Inuit* (Inuit) a 'real' man; an Eskimo

kaffemik (Inuit) coffee; a coffee party
kamik (Inuit) sealskin boot
kusanaq (Inuit) beautiful, handsome

mamarpoq (Inuit) good

massakkut (Inuit) at once, right away

mattak (Inuit) edible whale skin

meeqqat (Inuit) children; singular, *meeraq*

Mikilissuaq (Inuit) 'Big Michel' (Kpomassie)

mitaartoq (Inuit) a young man or woman disguised as a spirit, clad in furs; plural, *mitaartut*

naamik (Inuit) no

Nagan (Mina) respectful form of address to a mother senior to one's own

Nuuk (Inuit) the settlement of Godthåb

Paamiut (Inuit) the settlement of Frederikshåb

paarnat (Inuit) a small edible berry, a kind of blueberry

palasi (Inuit) Protestant pastor

pisiniarfik (Inuit) general store, supermarket

puisi (Inuit) seal

Qallunaaq (Inuit) usually translated as 'white man', but also applied to Kpomassie, so more correctly 'foreigner'

Qaqortoq (Inuit) the settlement of Julianehåb

qassi (Inuit) ceremonial meeting house; whale jaws are used in the construction

qeeraq (Inuit) the wolffish or sea wolf, a carnivorous fish

qimmeq (Inuit) husky; plural *qimmit*

qujanaq (Inuit) thank you; *kaffimut qujanaq*, thanks for the coffee, is endlessly used in Greenland!

Sakpatê (Mina) the earth goddess

tassa! (Inuit) be quiet, shut up!

timmisartoq (Inuit) helicopter

umiaq (Inuit) sealskin boat, used by women

Yovo (Mina) the whites